Sustainable Packaging Symposium 2012

Houston, Texas, USA
3-5 April 2012

ISBN: 978-1-5108-1768-5

Printed from e-media with permission by:

Curran Associates, Inc.
57 Morehouse Lane
Red Hook, NY 12571

CURRAN ASSOCIATES INC. proceedings.com

Some format issues inherent in the e-media version may also appear in this print version.

Printed by Curran Associates, Inc. (2016)

For permission requests, please contact AIChE
at the address below.

AIChE
120 Wall Street, FL 23
New York, NY 10005-4020

Phone: (800) 242-4363
Fax: (203) 775-5177

www.aiche.org

Additional copies of this publication are available from:

Curran Associates, Inc.
57 Morehouse Lane
Red Hook, NY 12571 USA
Phone: 845-758-0400
Fax: 845-758-2634
Email: curran@proceedings.com
Web: www.proceedings.com

TABLE OF CONTENTS

ecomagination℠

ecoassessment and the supply chain
Life Cycle Management at GE

Bill Flanagan
GE Global Research

Sustainable Packaging Symposium
Houston, TX
April 4, 2012

imagination at work

Life Cycle Assessment (LCA)

Assess overall environmental impact throughout a product or service's life cycle:

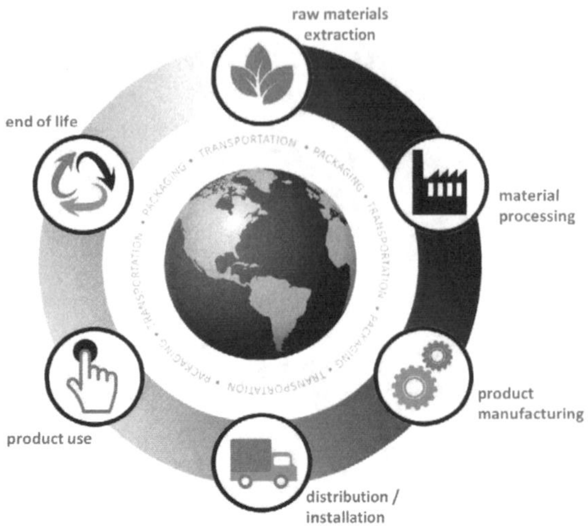

... including packaging and transportation

Quantify ins & outs

Energy
Materials
Emissions

- Fossil Fuels
- Hazardous Wastes
- Minerals
- Land Use
- Climate Change
- Ozone Depletion
- Respiratory Organics
- Resp. Inorganics
- Radiation
- Acidification
- Eutrophication
- Ecotoxicity
- Cumulative Energy Demand

ISO guidelines 14040-44

Increasing drivers in the market

Customer

- Increasing demand for products, services, and solutions that provide economic value with lower environmental impact
- Growing environmental disclosure requests
- More supply chain sustainability initiatives

Regulator

- Increasing international pressures for sustainability performance & disclosure
- Strengthening environmentally focused public procurement programs
- Stricter green marketing regulations

Competitor

- Growing availability of environmentally preferred products (EPP) and environmental product declarations (EPD)
- Increasing environmental disclosure via CDP and other standards
- Opportunity for thought leadership, market influence

Strong trends → Issues and Opportunities

Business-driven application of LCA

direct and indirect value

eco Product Innovation
LCA a key element of environmentally conscious product design
(but not the only element)

Commercial
(1) Ability to deliver complex environmental messaging;
(2) Ability to compete for bids requiring LCA / carbon footprint

Business Strategy
Identify strategic business opportunities

Due Diligence / Risk Management
Identifying and addressing potential perceptual and business risks

Reputation
(1) Enhancing corporate reputation and eco brand value;
(2) Ensuring seat at environmental policy table

ecoassessment

center of excellence

A systematic way to assess environmental footprint of selected GE products

- Strategy and vision
- Expertise and guidance
- Tools and processes
- Education and awareness
- Policy and advocacy

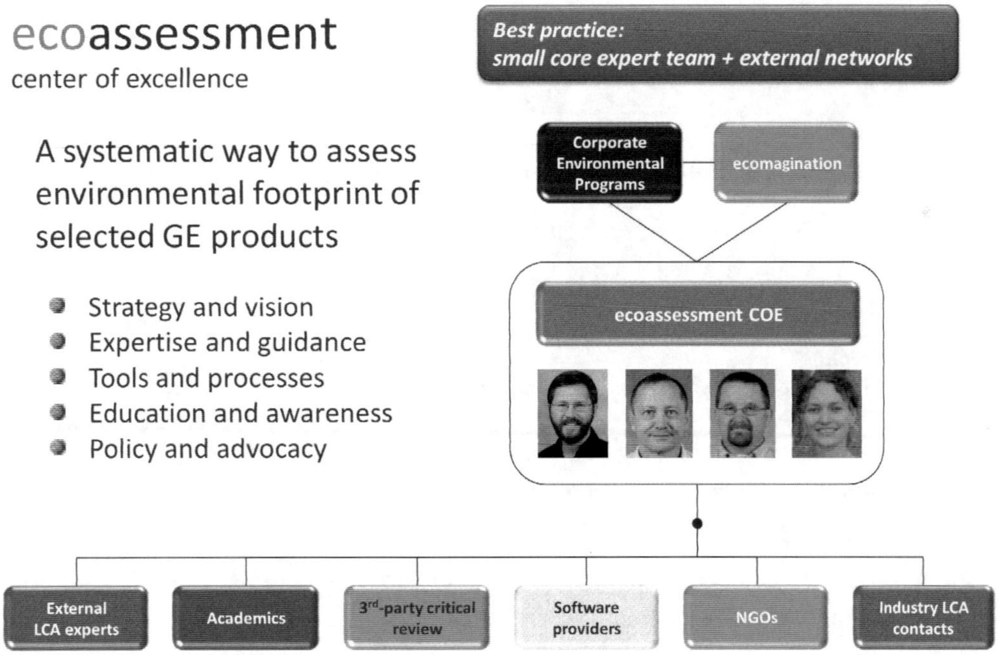

ecoassessment

Selected activities

Detailed LCA studies (examples)

- GE Healthcare: ReadyToProcess™ single-use process technology for biopharmaceutical manufacturing
- GE Energy: 2.5xl wind turbine
- GE Energy: Thin Film Solar
- GE Transportation: Durathon™ Battery

Screening/Streamlined LCA studies (examples)

- Appliances, Locomotives, Medical Equipment, Remanufacturing, Recycling, etc.

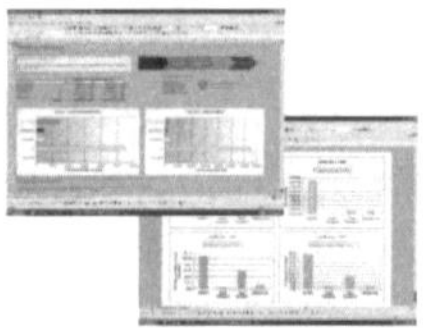
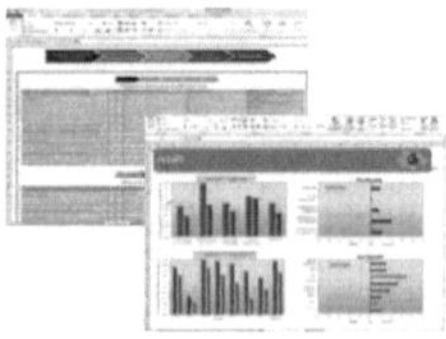

Tools development

- Screening LCA Tool
- Environmental Product LCM Tool (qualitative)
- Application-specific tools (e.g., wind, solar)

Ecomagination

- Ecomagination product certification support
- Life Cycle Management

Customer engagement

- Data requests (both directions)
- Strategic engagement
- Collaborative opportunities

Credibility build

Method development

- Uncertainty / sensitivity analysis
- Advanced numerical analysis

Outreach

- University outreach (lectures)
- Conference short courses

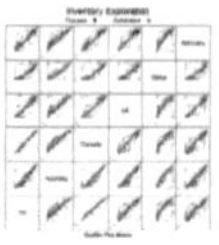
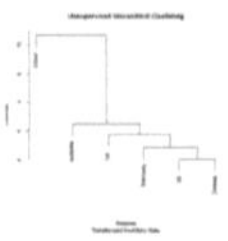

Life Cycle Management:

A Critical Strategy on the Path to Resource Efficiency

ecomagination℠

1. Life Cycle Management (LCM)

- A systematic approach to managing product-related environmental issues throughout a product's life cycle
- Can be broadly applied across a product portfolio
- Efficiently identifies opportunities for improvement; helps focus resources on areas that benefit from deeper analysis

2. Appropriate use of LCA

- Very powerful when applied in support of an LCM strategy (along with other tools and approaches)
- Ranges from qualitative → screening → streamlined → detailed
- Be selective and strategic: first determine the benefit

3. Goal of LCM (supported by LCA) is to…

- Identify the most promising opportunities for real improvement
- Create and maintain momentum toward real change
- Incentivize game-changing products, technologies, solutions

A Comprehensive LCM Approach

ecomagination[SM]

Capability build within businesses

Specialized expertise and tools require (ecoassessment COE)

1 Environmental Product LCM Tool
- Simple, quick, yet comprehensive
- Early awareness / directional guidance
- Broader than LCA

2 Screening LCA Tool
- High-level, data "lite"
- Hot spot identification

3 Streamlined LCA
- More detailed
- Suitable for internal use

4 Extensive LCA
- Very detailed, with sensitivity and uncertainty analyses
- Required for support of external claims
- ISO 14040-44
- 3rd-party panel review required for external claims involving comparative assertions

Strategic down-select → maximum benefit

Environmental LCM Tool

ecomagination[SM]

Simple, quick, yet comprehensive
Provides early awareness:
-opportunities & issues
Provides directional guidance:
-where to seek answers and support
-link to existing tools, processes, and expertise
-when to dig deeper

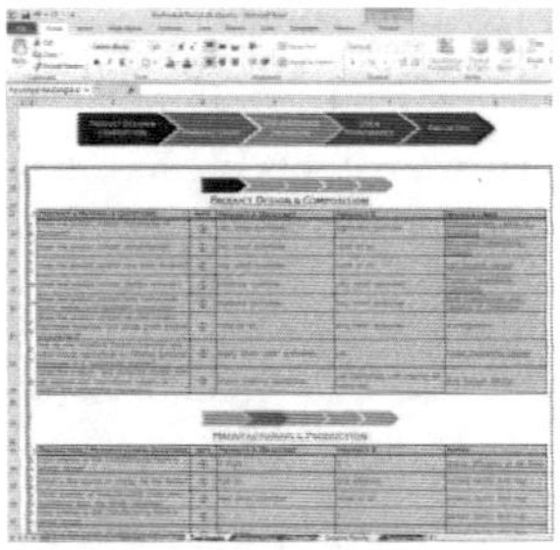

Qualitative inputs organized by life cycle stage
- Topic areas based on external stakeholder mapping
- Compare baseline vs. new product
- Compare product design alternatives

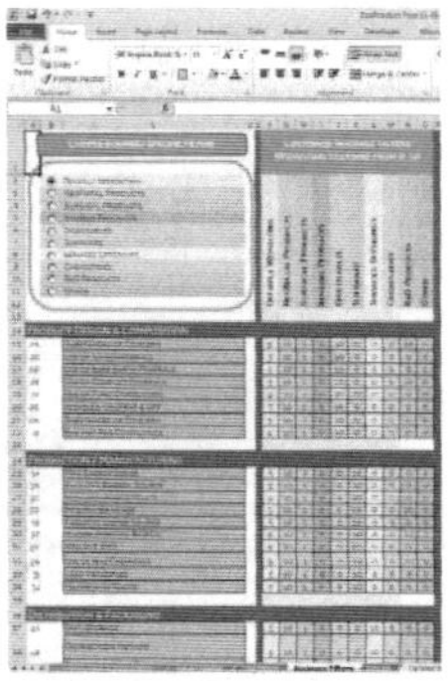

Customizable business context filters
- Calibrate tool responses based on relevancy per product category or business application

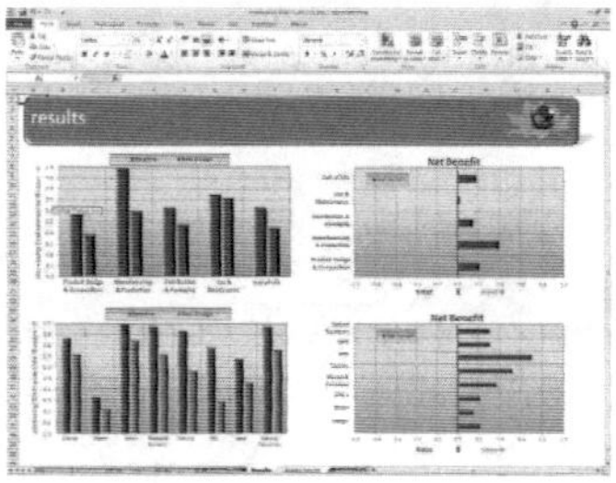

Graphical results
- Quickly identify opportunities or problem areas
- Results organized by:
  - (1) life cycle stage
  - (2) impact categories: energy, GHGs, water, wastes & emissions, toxicity, land, natural resources, EHS

ecoassessment coe
Angela Fisher & Bill Flanagan

Based on tool concept by Paul Holdredge (GEHC) + industry benchmarking

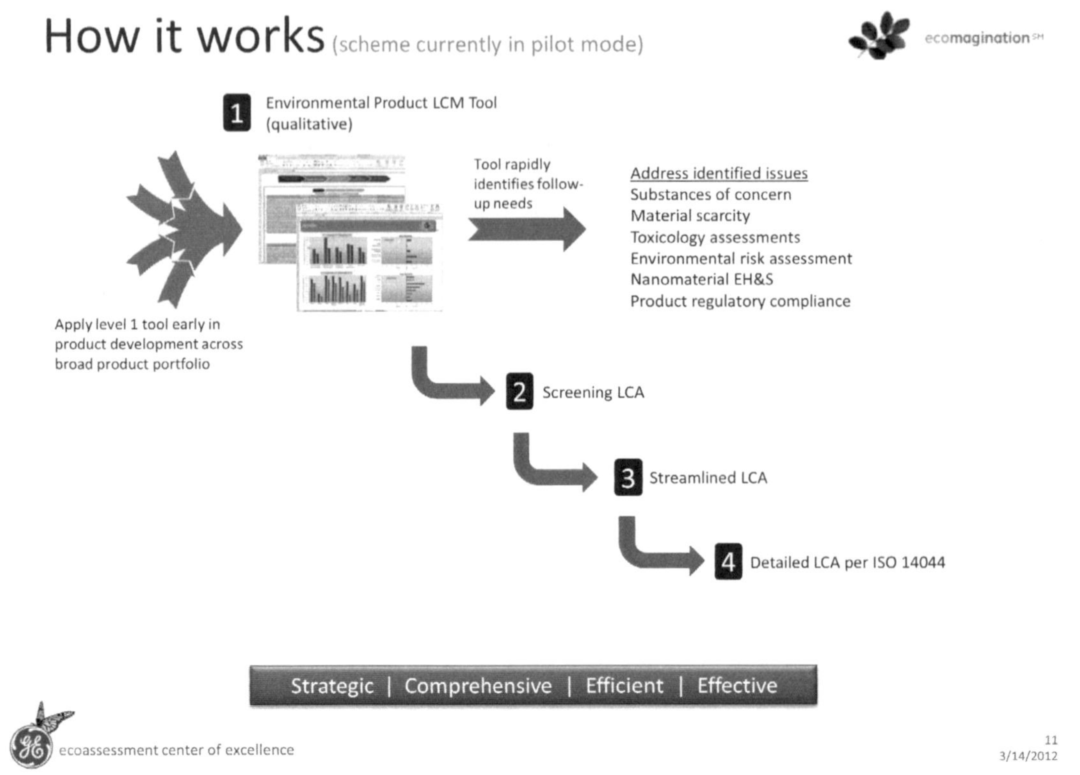
How it works (scheme currently in pilot mode)
ecomagination SM
1 Environmental Product LCM Tool (qualitative)
Tool rapidly identifies follow-up needs
Address identified issues
Substances of concern
Material scarcity
Toxicology assessments
Environmental risk assessment
Nanomaterial EH&S
Product regulatory compliance
Apply level 1 tool early in product development across broad product portfolio
2 Screening LCA
3 Streamlined LCA
4 Detailed LCA per ISO 14044
Strategic | Comprehensive | Efficient | Effective
ecoassessment center of excellence
11
3/14/2012

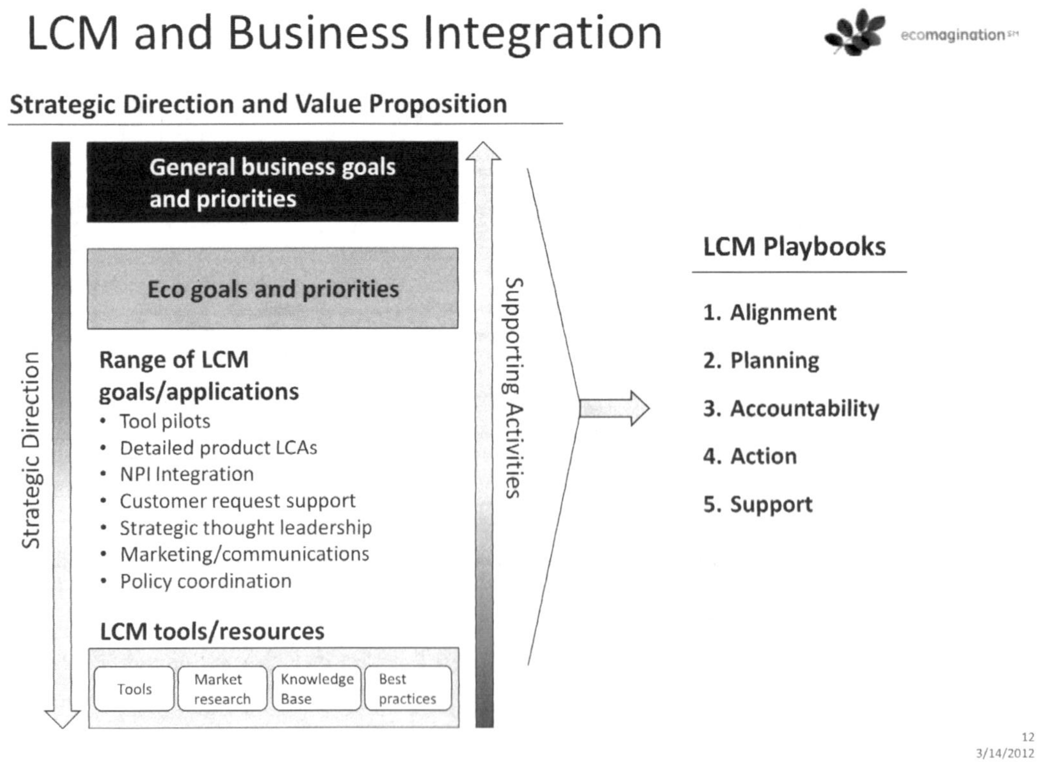
LCM and Business Integration
ecomagination SM
Strategic Direction and Value Proposition
General business goals and priorities
Eco goals and priorities
Range of LCM goals/applications
• Tool pilots
• Detailed product LCAs
• NPI Integration
• Customer request support
• Strategic thought leadership
• Marketing/communications
• Policy coordination
LCM tools/resources
Tools
Market research
Knowledge Base
Best practices
Strategic Direction
Supporting Activities
LCM Playbooks
1. Alignment
2. Planning
3. Accountability
4. Action
5. Support
12
3/14/2012

Enabling principles

ecomagination℠

Focus on value creation

- ✓ For any idea to thrive within industry, it must create value

Customize to business context

- ✓ No "one size fits all" tool or strategy
- ✓ Enhances relevance, uptake, ownership

Leverage power of innovative thinking

- ✓ Invite active engagement
- ✓ Great ideas can come from anywhere

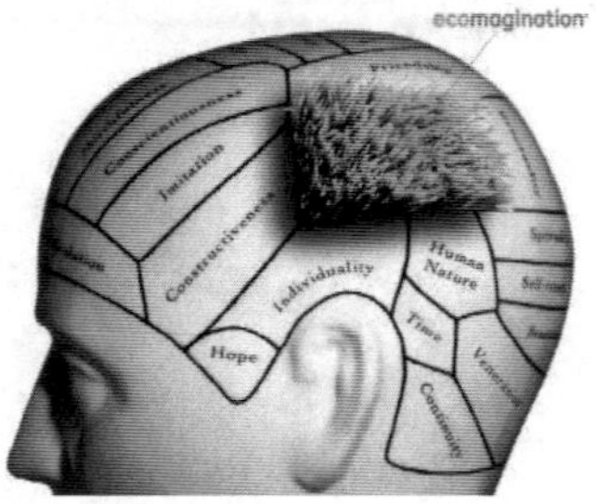

ecoassessment center of excellence

13
3/14/2012

Challenges

Pace of change

- ✓ Product development timelines require nimble approaches

Supply chains are dynamic

- ✓ Change occurs more quickly than data collection

Primary data collection is expensive

- ✓ Supply chain
- ✓ Internal within company
- ✓ Customer

Streamlined tools and strategies are essential

ecoassessment center of excellence

14
3/14/2012

medical packaging LCA example

ecoassessment center of excellence

15
3/14/2012

Packaging options for today's needs

Strong, growing demand
> 60% of units
sold are +*PLUS*PAK

+*PLUS*PAK™
(polymer bottle)

An eco**magination** product

Glass bottles & vials

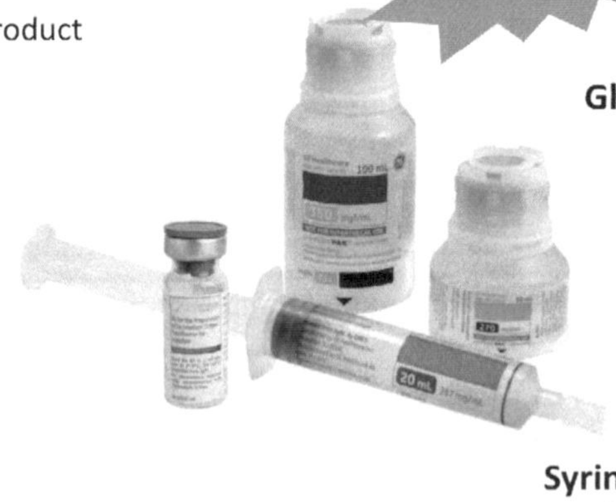

Syringes
Four hand-held syringes

Versatility – Workplace Safety - Efficiency

ecoassessment center of excellence

16
3/14/2012

+*PLUS*PAK (polymer bottle) packaging
for workplace safety and an Eco-friendly alternative

Feature	Glass Bottles	+PLUSPAK™ (polymer bottle)
Construction	Glass can break, posing sharps hazard	Durable, pharmaceutical-grade polymer construction
Stopper holder	Metal, which presents hazard of finger cuts from sharp edges	Easy twist-off cap with NO metal
Disposal	Empties discarded in special sharps container	Dispose in regular waste disposal
Weight	70%-80% heavier than +*PLUS*PAK packaging	Compact size, lighter weight
Accountability/ Record-keeping		Peel-off tracking labels, latex-free stopper

Goal and Scope

Goal:

To assess and compare the life cycle environmental impacts associated with +PLUSPAK and glass packages used for delivering contrast agent for X-ray and MRI procedures.

Functional unit:

Packaging of contrast agent required to deliver one dose of 96 ml to a patient for an X-ray procedure.

Packaging of contrast agent required to deliver one dose of 17 ml to a patient for an MRI procedure.

LCA consultant for this project: EarthShift LLC
Harnoor Dhaliwal, Jon Griesser, Lise Laurin

System Boundaries

+PLUSPAK bottle

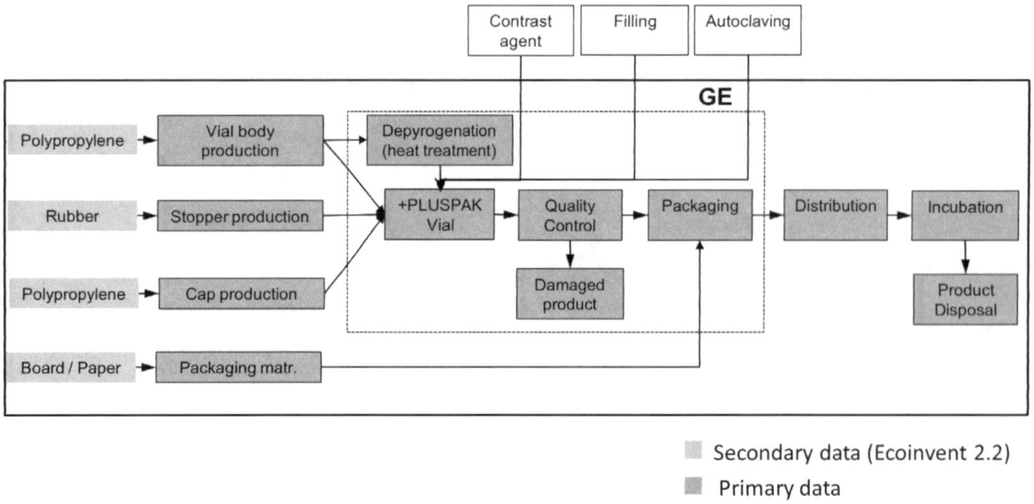

System Boundaries

Glass bottle

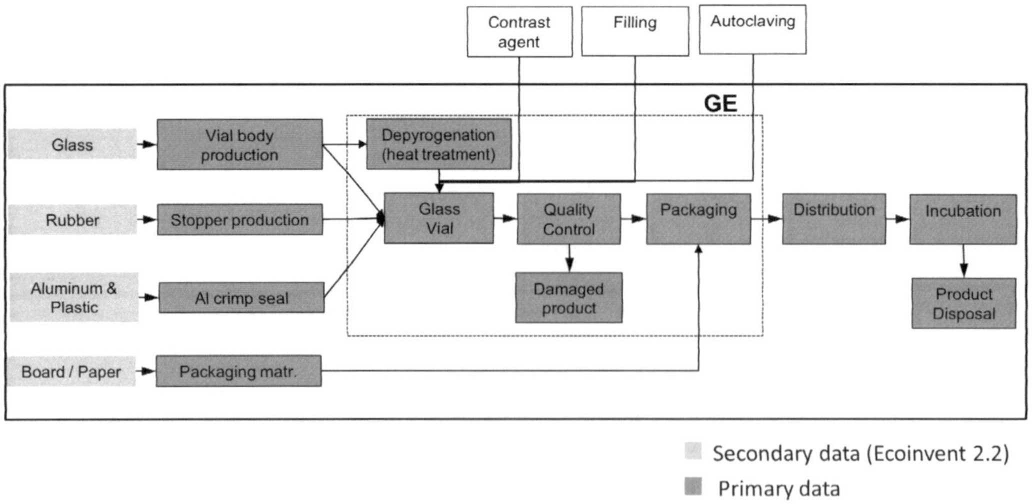

Disclaimer: Study is in the process of undergoing 3rd-party critical panel review. Results are preliminary and may be subject to adjustment.

Comparison

+PLUSPAK vs. Glass

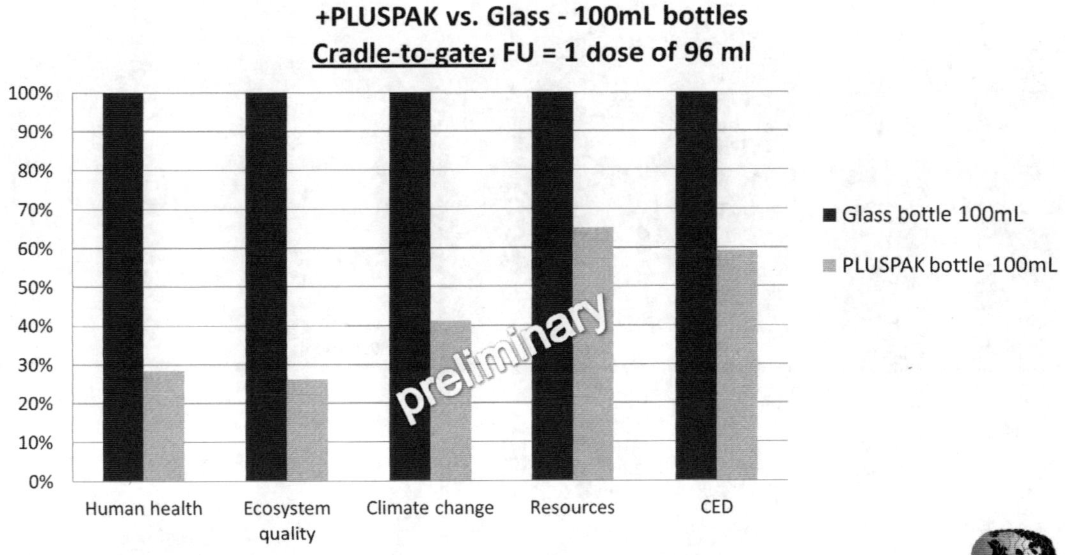

LCIA method: IMPACT 2002+ & Cumulative Energy Demand method

EarthShift
Building pathways to sustainability

Disclaimer: Study is in the process of undergoing 3rd-party critical panel review. Results are preliminary and may be subject to adjustment.

Comparison

+PLUSPAK vs. Glass

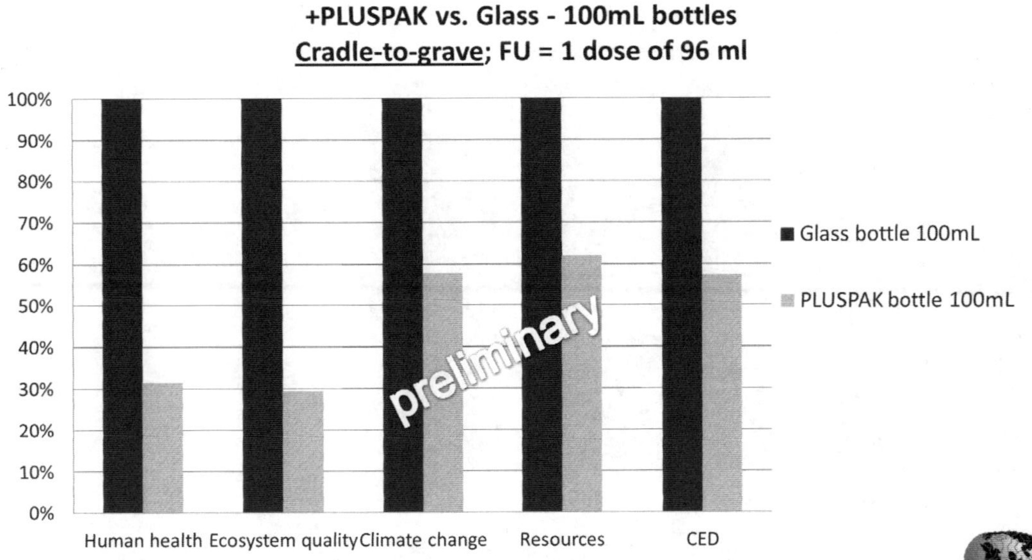

LCIA method: IMPACT 2002+ & Cumulative Energy Demand method

EarthShift
Building pathways to sustainability

Disclaimer: Study is in the process of undergoing 3rd-party critical panel review. Results are preliminary and may be subject to adjustment.

Contribution analysis

+PLUSPAK bottle: cradle-to-grave

100 mL +PLUSPAK bottle
FU = 1 dose of 96 ml

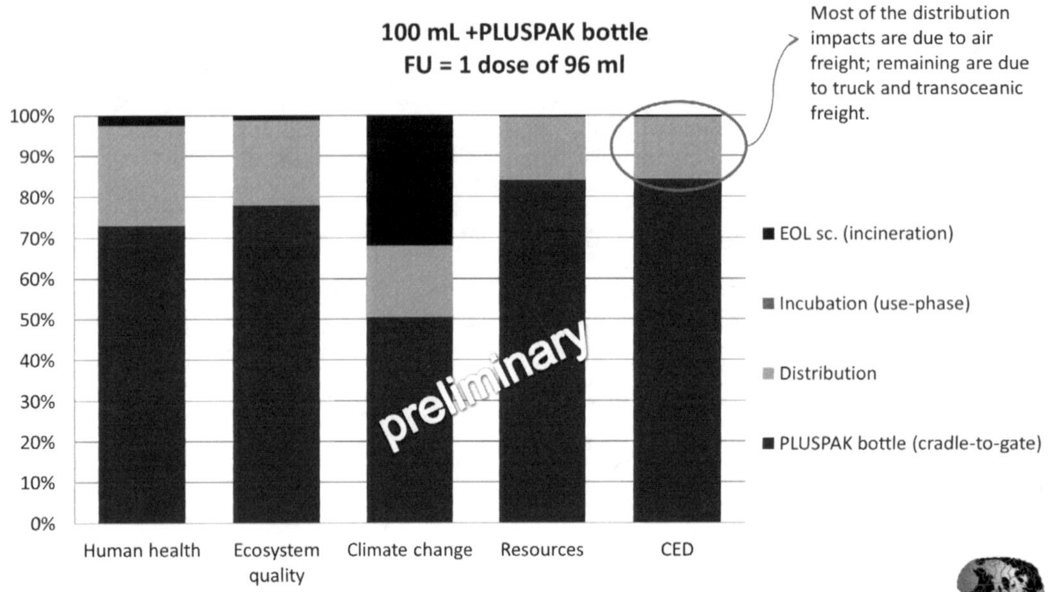

LCIA method: IMPACT 2002+ & Cumulative Energy Demand method

EarthShift
Building pathways to sustainability

Disclaimer: Study is in the process of undergoing 3rd-party critical panel review. Results are preliminary and may be subject to adjustment.

Contribution analysis

Glass bottle: cradle-to-grave

100 mL glass bottle
FU = 1 dose of 96 ml

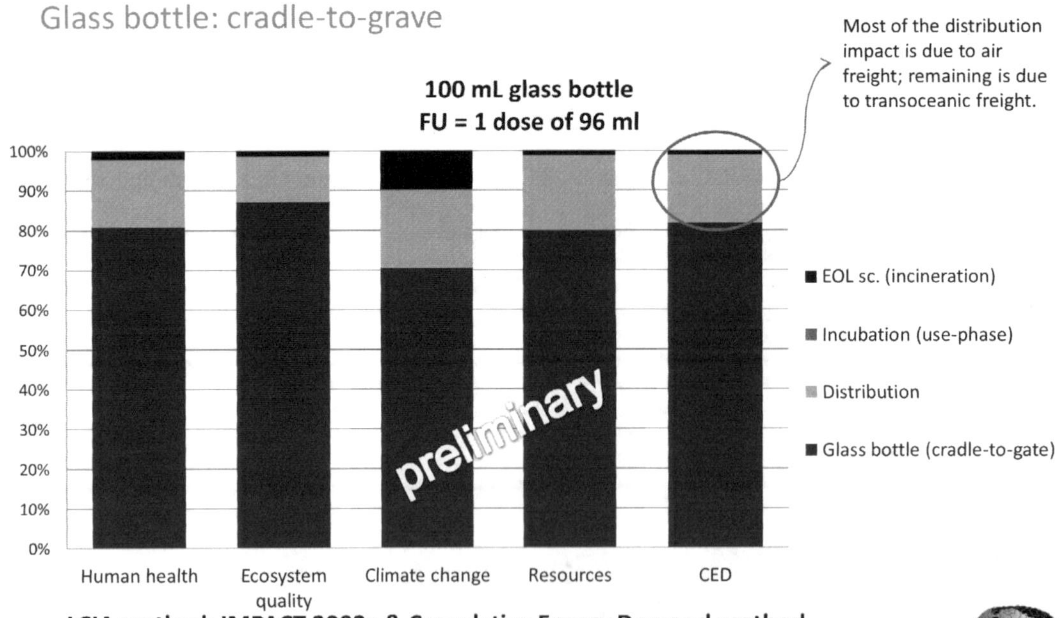

LCIA method: IMPACT 2002+ & Cumulative Energy Demand method

EarthShift
Building pathways to sustainability

Contribution analysis

+PLUSPAK bottle: cradle-to-gate

Disclaimer: Study is in the process of undergoing 3rd-party critical panel review. Results are preliminary and may be subject to adjustment.

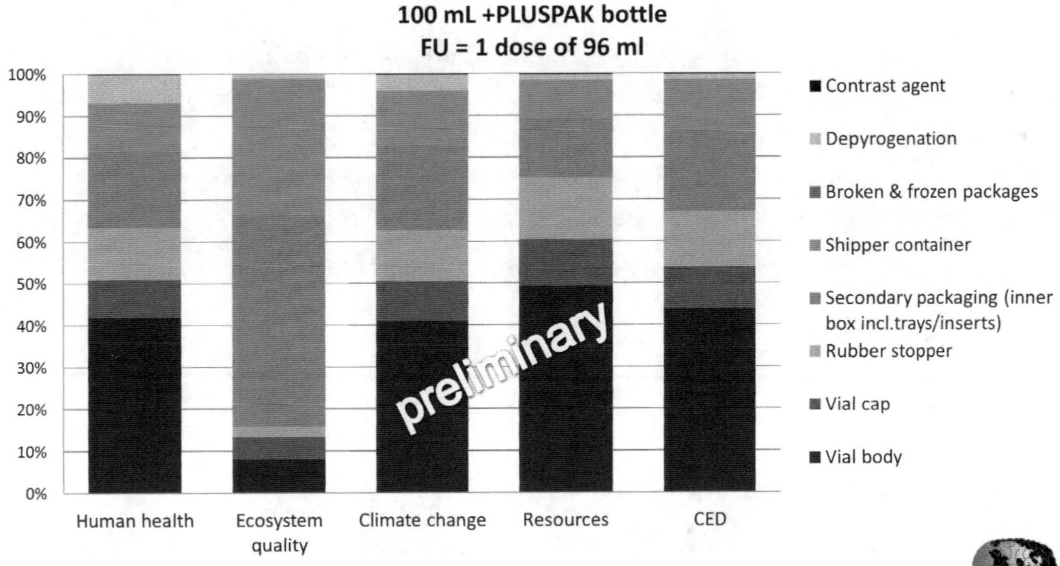

LCIA method: IMPACT 2002+ & Cumulative Energy Demand method

EarthShift
Building pathways to sustainability

Contribution analysis

Glass bottle: cradle-to-gate

Disclaimer: Study is in the process of undergoing 3rd-party critical panel review. Results are preliminary and may be subject to adjustment.

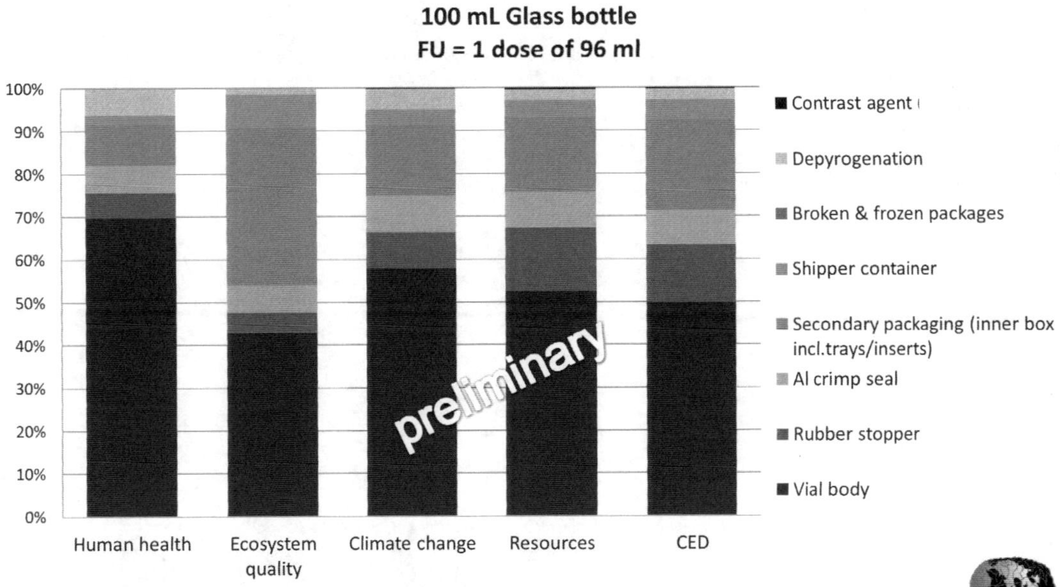

LCIA method: IMPACT 2002+ & Cumulative Energy Demand method

EarthShift
Building pathways to sustainability

Preliminary findings

Disclaimer: Study is in the process of undergoing 3rd-party critical panel review. Results are preliminary and may be subject to adjustment.

- Bottle vials are one of the most dominant sources of impact for +PLUSPAK and glass bottles;
- Secondary package and shipper container are important contributors to impacts;
- Air freight contributes to majority of the distribution related impacts;
- The contrast agent lost due to the breakage of the bottles contributes approximately 1% or less to the impacts

LCA can be a very powerful approach to understanding packaging impacts in the context of their application

ecoassessment
center of excellence

Bill Flanagan
flanagan@ge.com

Angela Fisher
fishera@ge.com

Matt Pietrzykowski
pietrzyk@ge.com

Ron Wroczynski
wroczynski@ge.com

PE INTERNATIONAL
EXPERTS IN SUSTAINABILITY
FiveWinds
STRATEGIC CONSULTING

Leadership in the Sustainable Supply Chain

Laura Flanigan, Senior Consultant
Sustainable Packaging Symposium
April 4, 2012

Progress Continues towards Sustainable Packaging

PE INTERNATIONAL
EXPERTS IN SUSTAINABILITY
FiveWinds
STRATEGIC CONSULTING

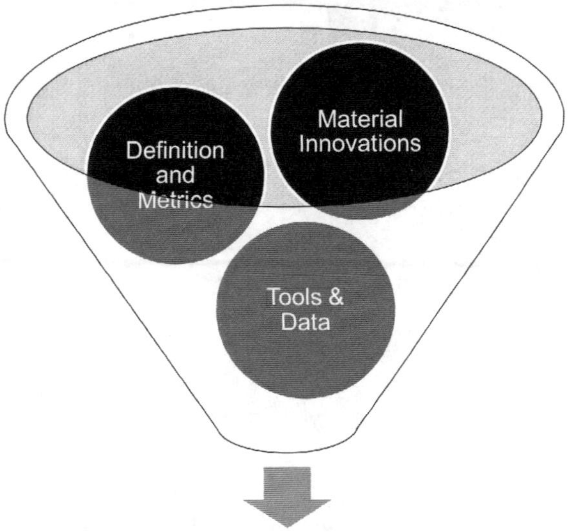

Sustainable Packaging ?

Problem: Multiple Voices, Multiple Priorities

PE INTERNATIONAL EXPERTS IN SUSTAINABILITY & FiveWinds STRATEGIC CONSULTING

- Aligning with overall business
- Managing overall product life cycle

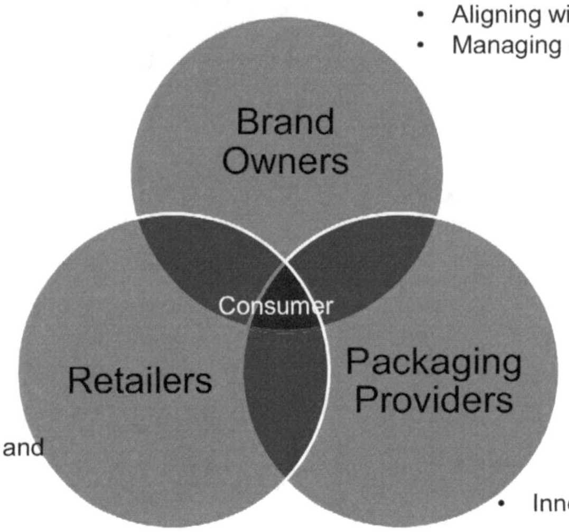

- Large scope of sustainability impact and influence

- Innovation and differentiation

Characteristics of Leadership

PE INTERNATIONAL EXPERTS IN SUSTAINABILITY & FiveWinds STRATEGIC CONSULTING

Use the right tool

Communicate

PE INTERNATIONAL
EXPERTS IN SUSTAINABILITY
& FiveWinds
STRATEGIC CONSULTING
Your partner in corporate and product sustainability performance
Laura Flanigan, Senior Consultant
l.flanigan@fivewinds.com

Ahold USA

Responsible Retailing

Sustainable Packaging Symposium 2012

Harriet Hentges

Vice President, Corporate Responsibility & Sustainability

Ahold GIANT Food Stores MARTIN'S Giant Stop&Shop Peapod

Leadership

- Vision
- Process
- Champion attributes

Ahold USA

Success comes from

- Clear targets
- Clear definitions
- Clear road map

Ahold
USA

Effectively connecting

- Connect the dots: relate what you have to offer to retailer's aspirations and goals
- Solve for the priorities
- Help solve for competing goals
- Understand how category manager/buyer is rewarded
- Grasp the nature of retailing

Ahold
USA

Putting the pieces together to create a sustainable picture

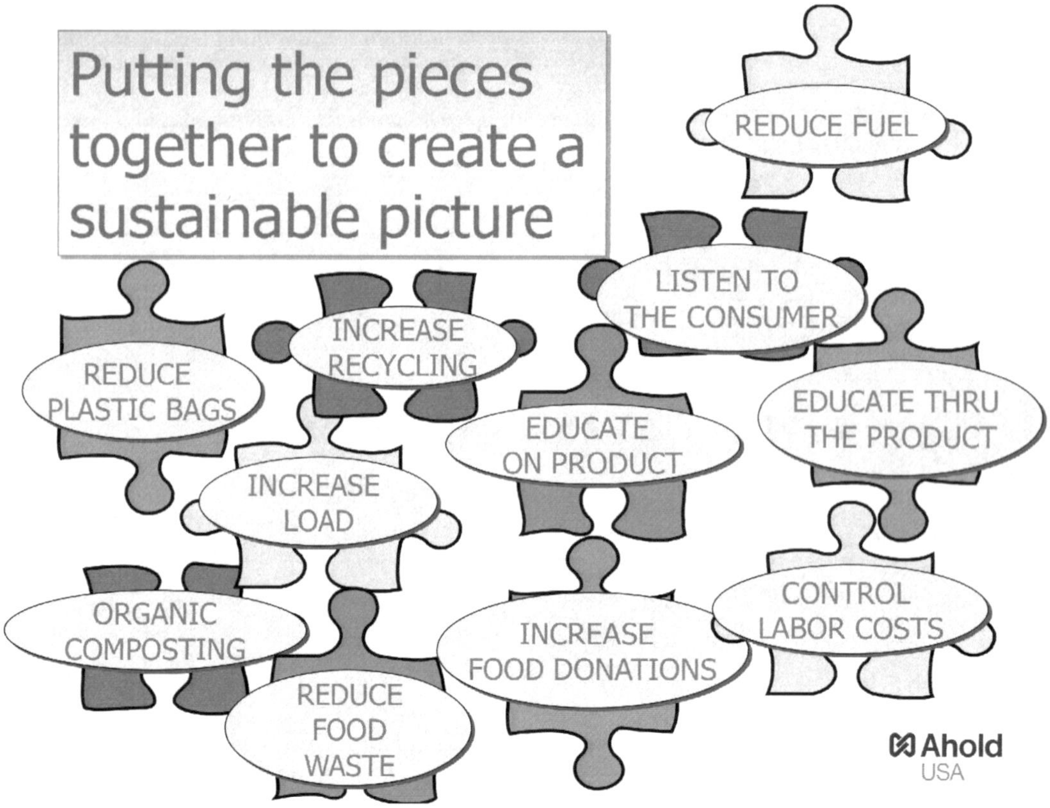

Ahold USA

HEALTHY CRITERIA

More:

Fiber

Protein

Vitamin A

Vitamin C

Calcium

Iron

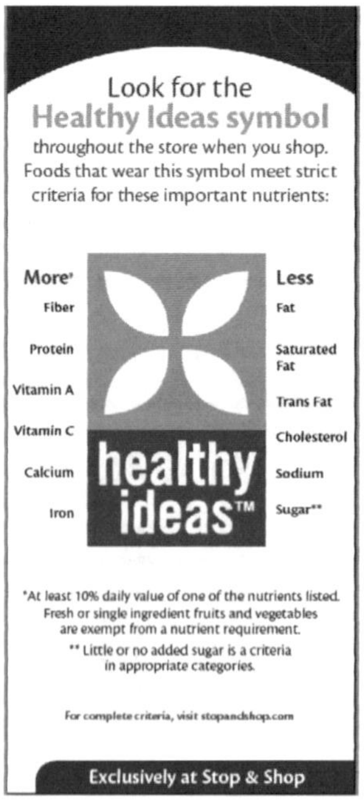

Less:

Fat

Saturated Fat

Trans Fat

Cholesterol

Sodium

Sugar

Ahold USA

Challenges

- Senior management buy-in & support
- Technical complexity and advances
- Sustainability cannot be a burden
  - Remember how category managers/buyers are measured
  - Make it **Easy**
  - Make it **Simple**
  - Make it **Relevant**
- Find the **Sweet Spot** – where business and sustainability goals intersect

Ahold USA

Business Problem: Product on Top 10 of damage report

Action

- Sourced / tested a high strength woven laminate to improve strength, reduce damages
- Chose a single polymer vs. a four-ply paper bag
- Puncture, probe and tear resistance increased

Ahold USA

Sweet Spot

Business Benefits

- Reduced in-transit damage and unsalable goods over 50% with over $650K annual savings
- More controllable surface; easier to palletize; doesn't move in transport
- **No longer in top 10 on damage report!**

Environmental Benefits

- Reduced packaging weight by more than 50%
- Post use solid waste reduction estimated to be over 1400 tons per year
- Fuel & number of trees saved calculated

Ahold USA

Johnson & Johnson

FAMILY OF CONSUMER COMPANIES

Aspiration

How do we define Healthy Future?

Healthy Future means fulfilling our Credo responsibilities so people, planet and business can all live in good health, today and tomorrow.

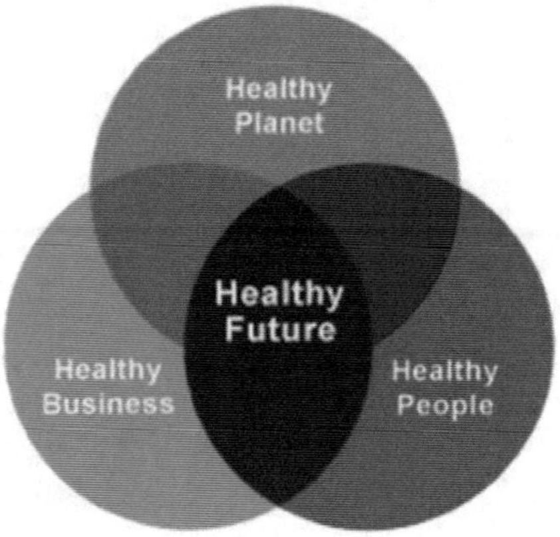

In the Consumer Group, we have a unique set of aspirations:

Healthy People: Health & wellbeing for all

- Touch more lives every day
- Keep our employees safe & well
- Empower people to stay well

Healthy Planet: A world without waste

- Use fewer natural resources
- Use more renewable resources
- Design for the next life

Healthy Business: Most trusted brands

- Share our stories
- Be the partner of choice
- Inspire people to make a difference

Johnson & Johnson FAMILY OF CONSUMER COMPANIES

Communication

Highlights: Employee Engagement

Healthy Future Communications

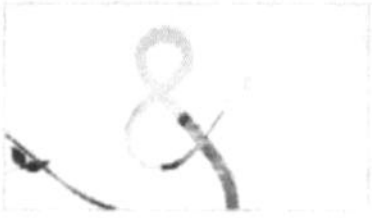

Johnson & Johnson FAMILY OF CONSUMER COMPANIES

Spreading the Word

BRAINSTORM GREEN Johnson & Johnson TEDxJNJ Yale

Talking about a Healthy Future

Johnson & Johnson FAMILY OF CONSUMER COMPANIES

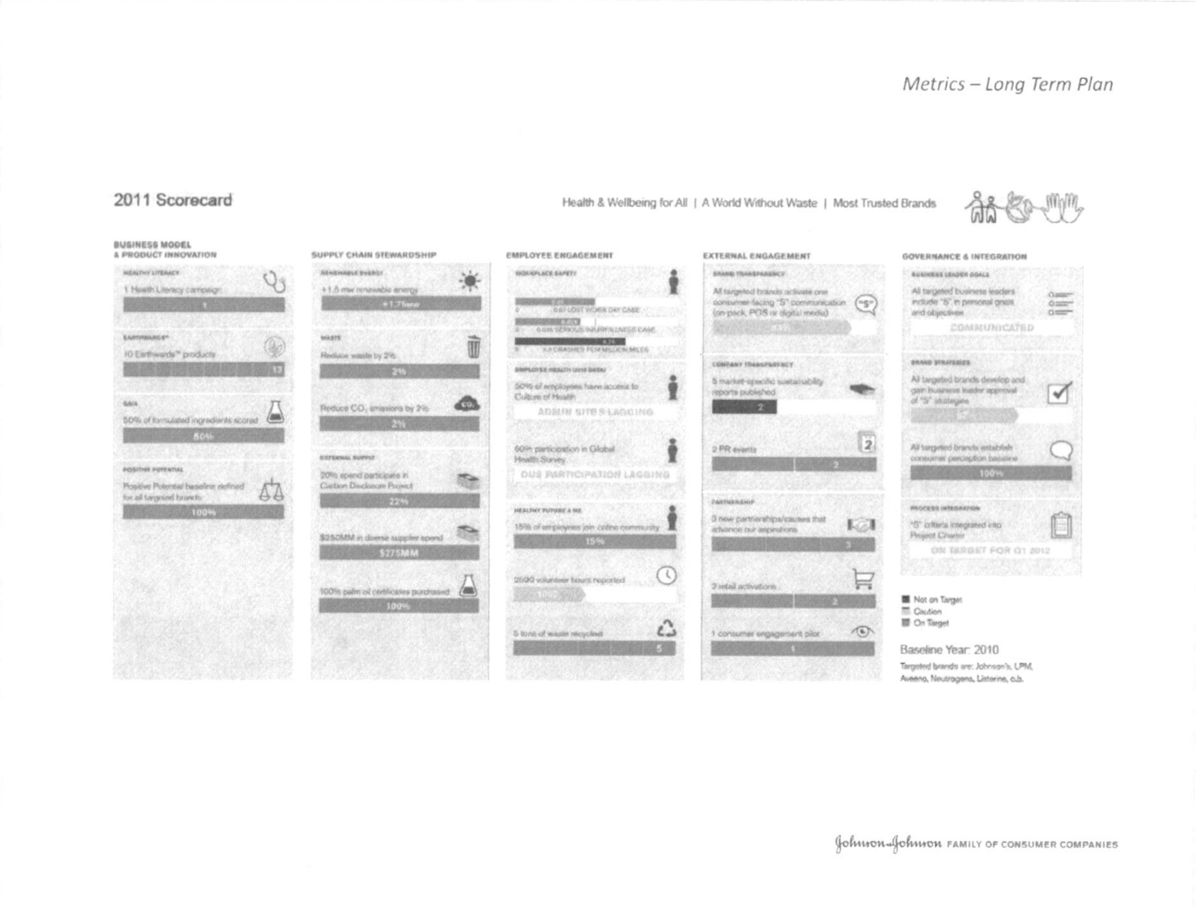
Metrics – Long Term Plan
2011 Scorecard
Health & Wellbeing for All | A World Without Waste | Most Trusted Brands
BUSINESS MODEL & PRODUCT INNOVATION
1 Health Literacy campaign
1
10 Earthwards™ products
13
50% of formulated ingredients scored
50%
Positive Potential baseline defined for all targeted brands
100%
SUPPLY CHAIN STEWARDSHIP
+1.5 mw renewable energy
+1.75mw
Reduce waste by 2%
2%
Reduce CO_2 emissions by 2%
2%
20% spend participate in Carbon Disclosure Project
22%
$250MM in diverse supplier spend
$275MM
100% palm oil certificates purchased
100%
EMPLOYEE ENGAGEMENT
LOST WORK DAY CASE
SERIOUS INJURY/ILLNESS CASE
CRASHES PER MILLION MILES
50% of employees have access to Culture of Health
ADMIN SITES LAGGING
60% participation in Global Health Survey
OUS PARTICIPATION LAGGING
15% of employees join online community
15%
2000 volunteer hours reported
5 tons of waste recycled
5
EXTERNAL ENGAGEMENT
All targeted brands activate one consumer-facing "S" communication (on-pack, POS or digital media)
5 market-specific sustainability reports published
2
2 PR events
2
3 new partnerships/causes that advance our aspirations
3
3 retail activations
2
1 consumer engagement pilot
1
GOVERNANCE & INTEGRATION
All targeted business leaders include "S" in personal goals and objectives
COMMUNICATED
All targeted brands develop and gain business leader approval of "S" strategies
All targeted brands establish consumer perception baseline
100%
"S" criteria integrated into Project Charter
ON TARGET FOR Q1 2012
Not on Target
Caution
On Target
Baseline Year: 2010
Targeted brands are: Johnson's, LPM, Aveeno, Neutrogena, Listerine, o.b.
Johnson-Johnson FAMILY OF CONSUMER COMPANIES

Our Tools

Our Tools
- Products

Highlights: Consumer Engagement

Neutrogena Naturals

FSC

The Nature Conservancy
Protecting nature. Preserving life.

- Safe ingredients that are naturally derived.
- Recycled paperboard folding cartons use 70% less water in processing than virgin paperboard.
- Partnerships with The Nature Conservancy and Forest Stewardship Council.

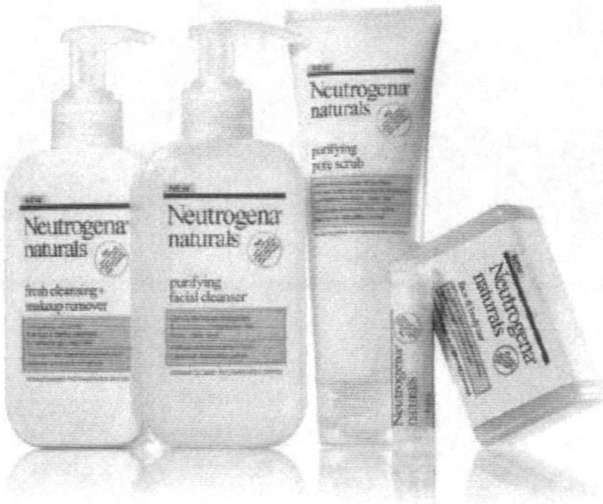

Johnson & Johnson FAMILY OF CONSUMER COMPANIES

Our Tools
- Supply Chain

Highlights: Supply Chain

Responsible Naturals

We partnered with WWF and completed a risk assessment of 11 natural ingredient feed stocks.

- What is the next palm oil?
- What are the opportunities to improve the sustainability of these supply chains?
- What are the opportunities to engage with stakeholders to make a difference?
- What are the stories we can tell our consumers?

Johnson & Johnson FAMILY OF CONSUMER COMPANIES

Highlights: Product Innovation

EARTHWARDS™ Recognized Products

earthwards

Johnson & Johnson FAMILY OF CONSUMER COMPANIES

Johnson & Johnson
FAMILY OF CONSUMER COMPANIES

The Importance of Packaging to the Sustainable Product Supply Chain

Beth Beloff
Beth Beloff & Associates

BB&A
Sustainability Strategies

Panel

- Jeff Meyers, Manager Sustainable Packaging Coca-Cola Refreshments
- Ryan L'Abbe, Vice President/General Manager Ice River Springs Water Co
- Joan Pierce, Executive Director AMERIPEN
- Beth Beloff, Principal, Beth Beloff & Associates; President, Bridges to Sustainability Institute

Why make supply chains more sustainable?

- To address greatest product lifecycle ***impacts***
- To create more ***efficient and effective*** supply chains
- To reduce ***costs***: water, carbon, energy…
- To reduce supply chain ***risks***
- To use ***purchasing criteria*** consistent with sustainability objectives
- To employ consistent and common ***measures*** of sustainability for all aspects of supply chain
- To collect consistent ***information*** from suppliers about their products and practices
- To assist suppliers in ***improving*** their sustainability efforts
- To improve ***product performance*** throughout the chain

BB&A
Sustainability Strategies

Context – Demands for Suppliers to demonstrate greater sustainability

- Four core elements
  - Product Performance
  - Product Composition
  - Product Life Cycle Management
  - Supplier Sustainability Practices

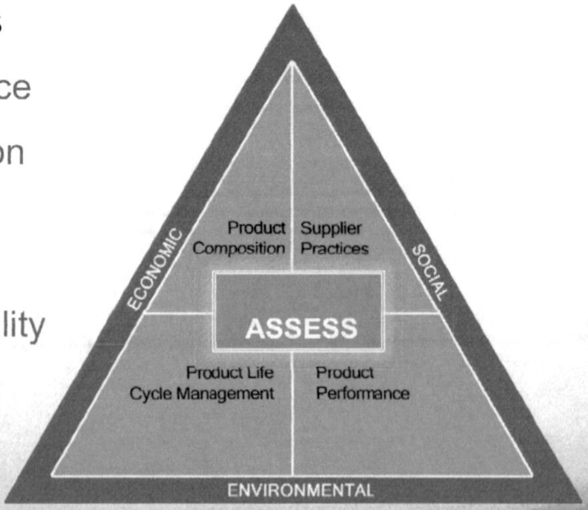

BB&A
Sustainability Strategies

Package Composition

- **Substances to Avoid** (CFCs, EPA Priority PBT substances, EPA Waste Min. Priority Chemicals, BPA, etc)
- **Environmental Attributes**:
  - Low Toxicity
  - Recycled materials
  - Bio-based materials
  - Material intensity reduction
  - Reusability, Recyclability, or Biodegradability
  - Sustainably produced bio-based materials (eg wood)
  - Embodied water, energy, carbon
  - Etc

BB&A
Sustainability Strategies

Package Lifecycle Management

- **Substances to Avoid** in Manufacturing
- **Key lifecycle impacts** have been identified and addressed
  - Energy use
  - Water use
  - GHG emissions
  - Solid waste
  - Hazardous waste
  - Wastewater effluent
  - Air emissions
  - Land use
  - End-of-life management

BB&A
Sustainability Strategies

Package Supplier Sustainability Practices

- **Corporate Governance**
  - Sustainability policy
  - Workplace, Human Rights, Ethics standards
  - Transparency in public reporting
- **Stakeholder relations / engagement**
- **Partnerships to address issues**
- **Environmental Management Systems**
- **Green procurement program**
- **Supply Chain management**
  - Supplies sourced manufactured in socially and environmentally responsible manner
  - Same level and social commitment of suppliers
  - Partner with supply chain players – close material recycle loop

BB&A
Sustainability Strategies

Package Performance in Use

- Meets **market criteria** and market **costs**
- Ensures that packaging does not compromise **safety and health** in use of product
- Protects product from **damage or loss**
- **Preserves** product
- Provides **information** regarding product attributes
- Enhances **durability** of product **/ extends shelf life** (eg electronic GPS/temperature gauge)
- Maintains **structural integrity** with **reduced material intensity**
- Helps reduce **carbon footprint**
- Ensures that materials are **reusable, recyclable, biodegradable** and are easily and efficiently recoverable
- **Unitization** for improvement on shelf and in transport

BB&A
Sustainability Strategies

How can Packaging Enhance Product/SC Performance?

- **Waste prevention**
  - Damage prevention – loss of product (eggs)
- **Durability**
- **Shelf life** extension
- **Accountability** - Identify chain of custody of product along its supply chain (microchips with GPS and temperature)
- **Improve Supply Chain** –
  - Reduce weight and compactness in transport
  - Unitize for improved transport
  - New uses at End-of-Life (recycle, biodegradability, use of packaging in reforestation, etc.)
  - Produce beneficial uses

BB&A
Sustainability Strategies

When assessment of impacts is extended to include product performance, different answers emerge……

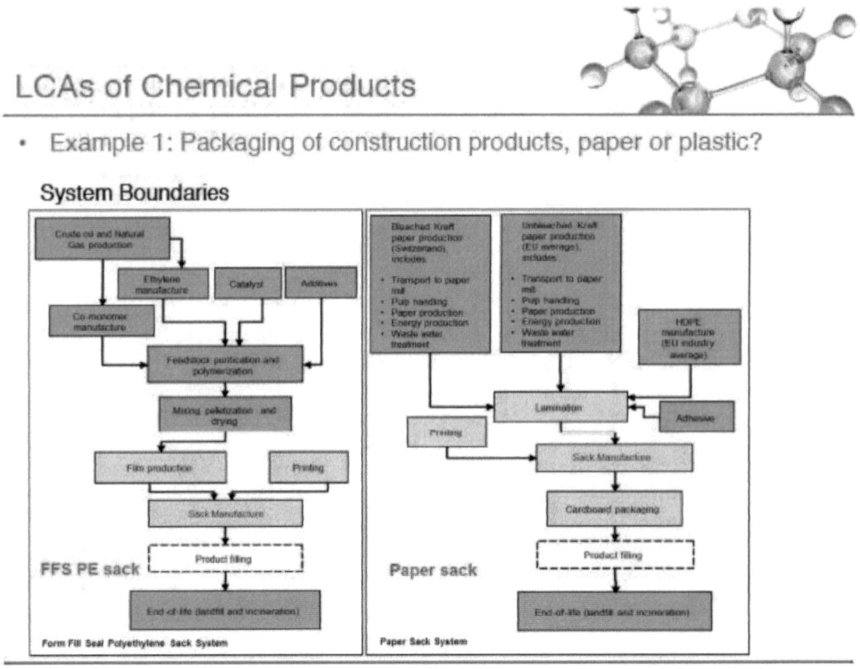

Source: Presentation by Abdelhadi Sahnoune, ExxonMobil Chemical Company for Carbon Management Technology Conference Feb 2012

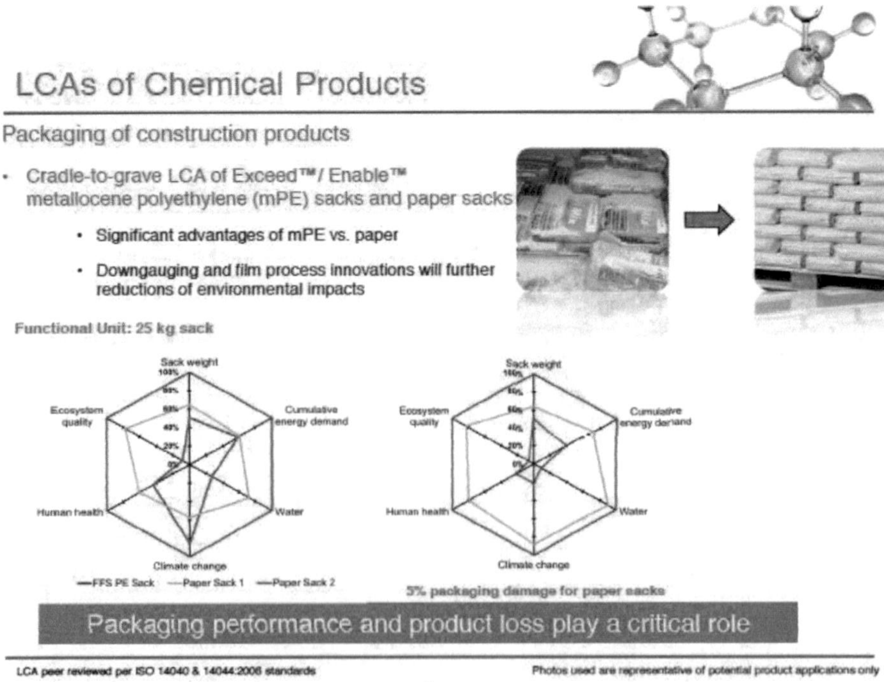

Source: Presentation by Abdelhadi Sahnoune, ExxonMobil Chemical Company for Carbon Management Technology Conference Feb 2012

The Life Box™

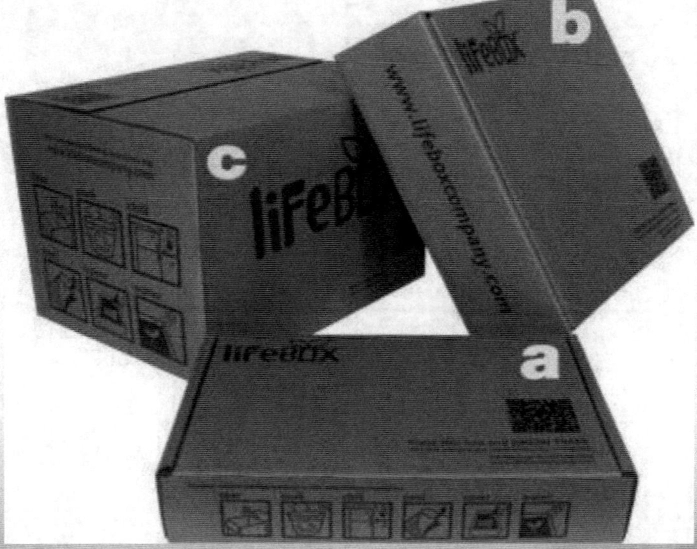

Thank you!

beth@bethbeloff.com

BB&A
Sustainability Strategies

The Importance of Packaging to the Sustainable Product Supply Chain

Sustainable Packaging Symposium
April 4th, 2012

Jeff Meyers
Manager, Sustainable Packaging

LIVE POSITIVELY Coca-Cola

Key Themes for Today

1. Supply chains are facing difficult sustainability challenges including rising costs
2. Improving the sustainability of packaging requires multiple approaches
3. Innovation is a critical element in delivering performance improvement and managing risk

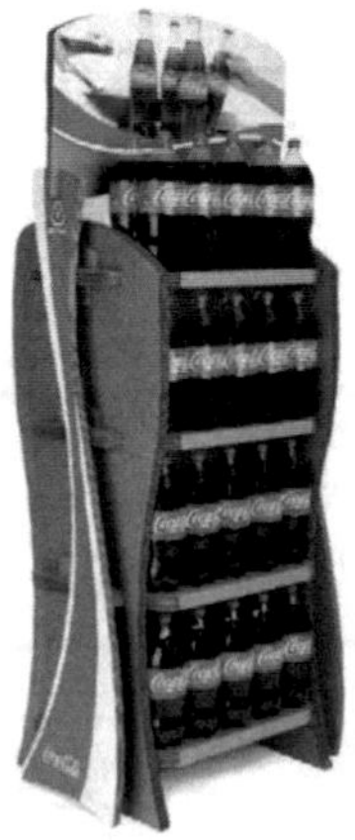

LIVE POSITIVELY Coca-Cola

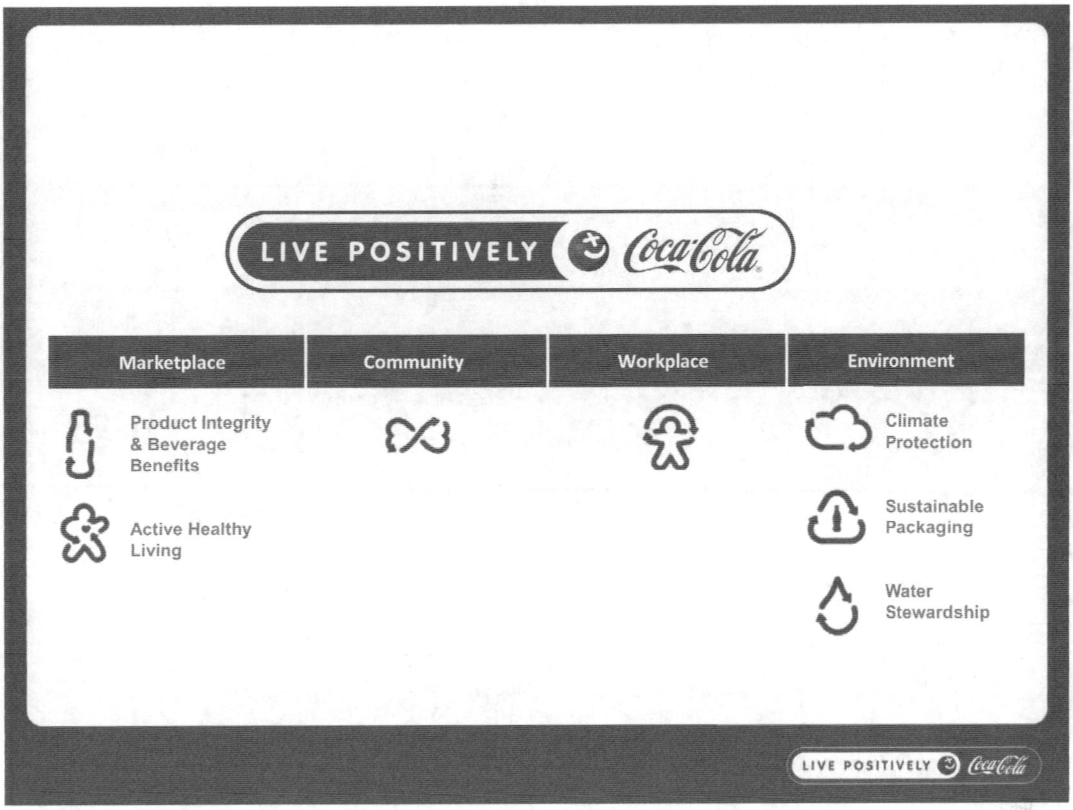

Supply Chain Trends

Some key challenges being faced by consumer goods companies regarding supply chains:

- Increased cost of carbon and water to manufacture and transport products
- Higher commodity prices for primary and secondary packaging
- Need for developing and improving reverse supply chains for product recycling
- Increased consumer interest in the overall sustainability of a supply chain

Innovation will be important to mitigate these impacts.

LIVE POSITIVELY Coca-Cola

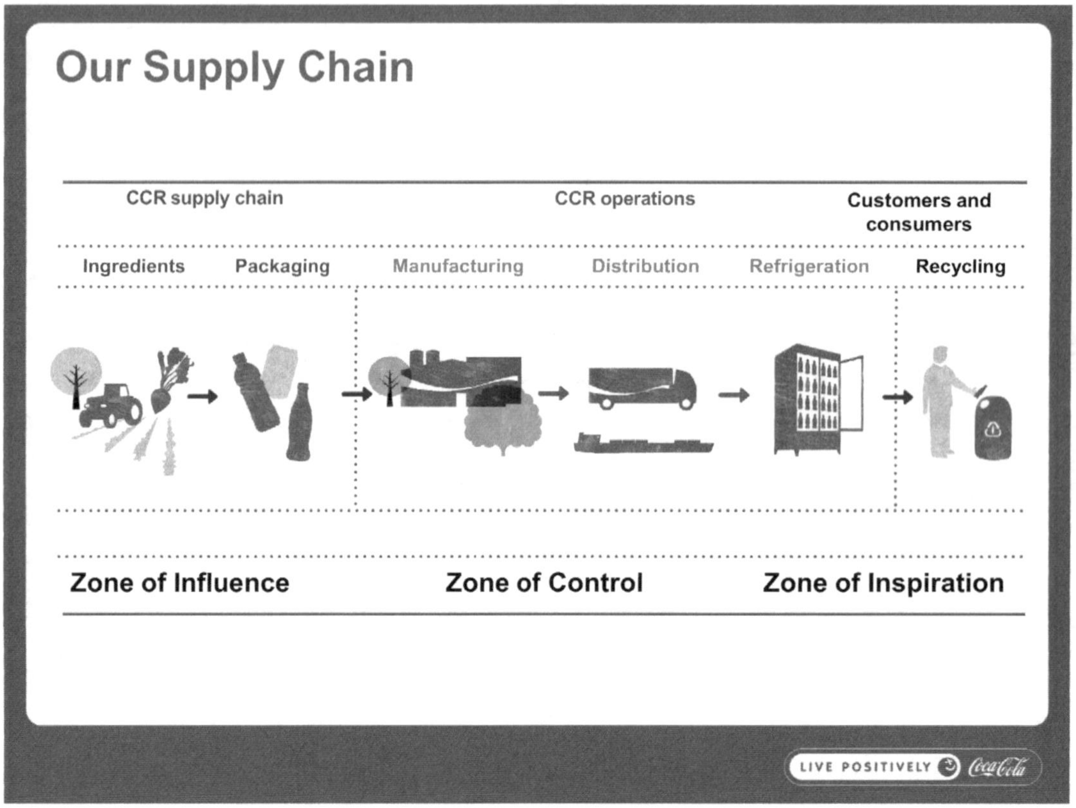
Our Supply Chain
CCR supply chain
CCR operations
Customers and consumers
Ingredients
Packaging
Manufacturing
Distribution
Refrigeration
Recycling
Zone of Influence
Zone of Control
Zone of Inspiration
LIVE POSITIVELY
Coca-Cola

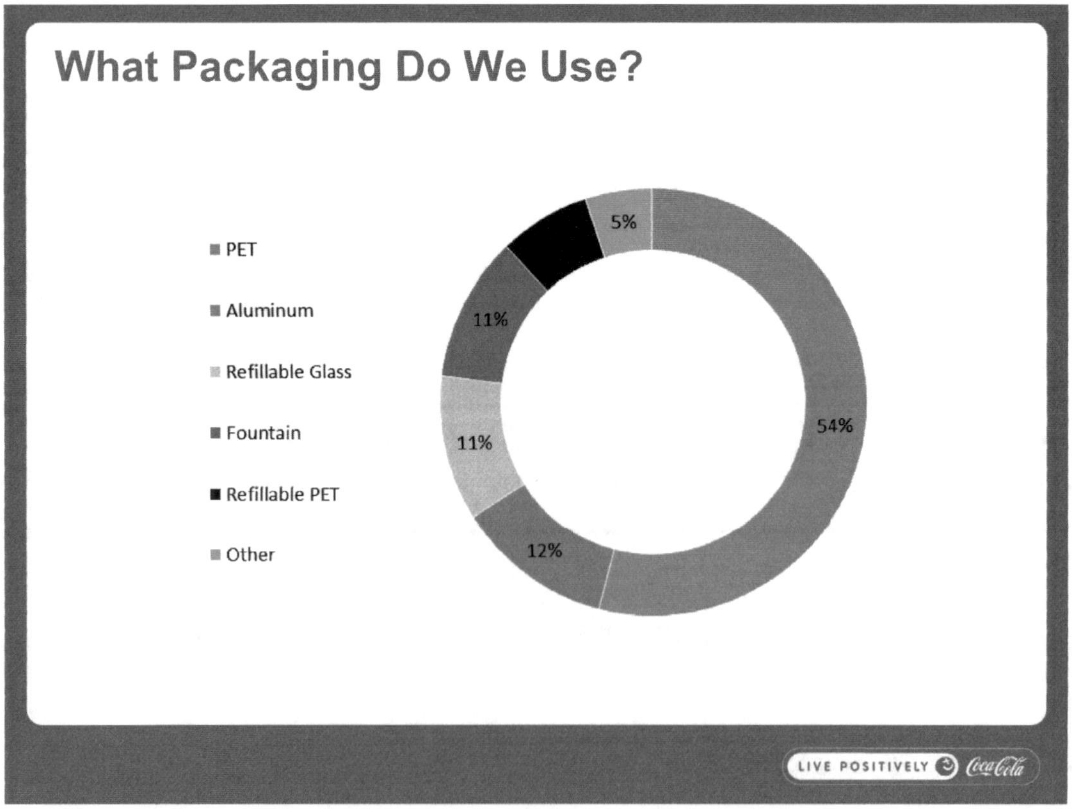
What Packaging Do We Use?
PET
Aluminum
Refillable Glass
Fountain
Refillable PET
Other
54%
12%
11%
11%
5%
LIVE POSITIVELY
Coca-Cola

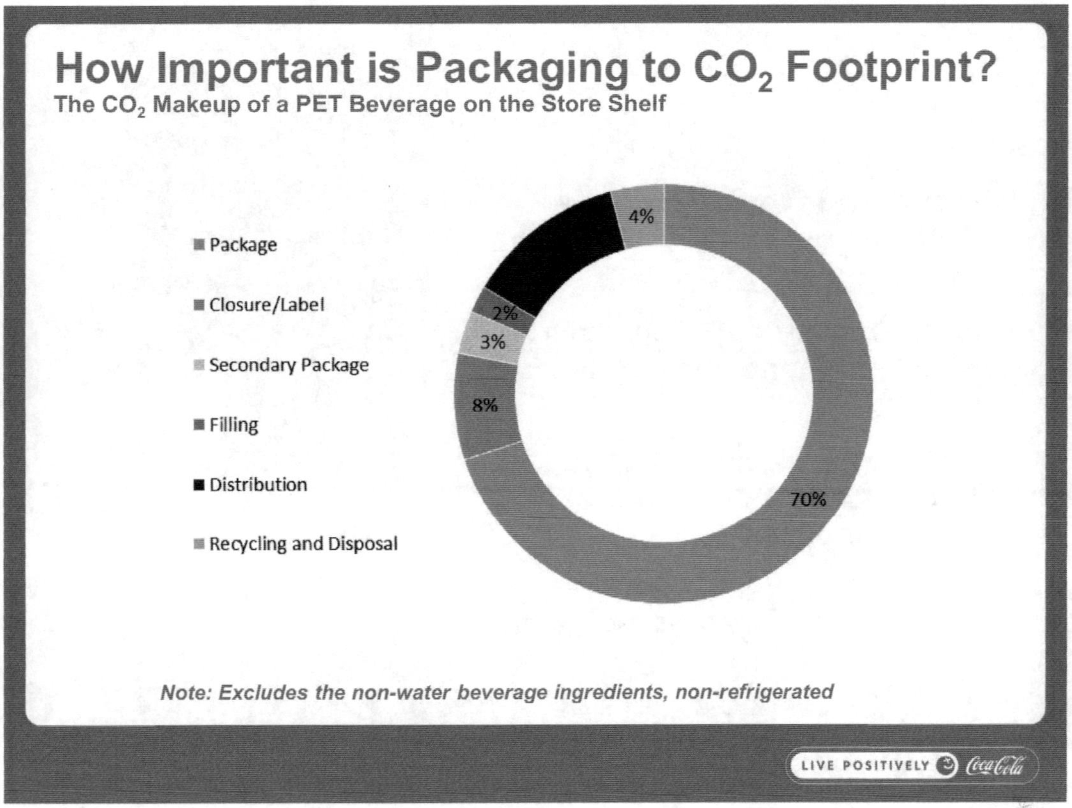
How Important is Packaging to CO_2 Footprint?
The CO_2 Makeup of a PET Beverage on the Store Shelf
Package
Closure/Label
Secondary Package
Filling
Distribution
Recycling and Disposal
4%
2%
3%
8%
70%
Note: Excludes the non-water beverage ingredients, non-refrigerated
LIVE POSITIVELY

Four Part Strategy to Improve
Reduce
Lightweighting
MORE THAN 50%
MORE THAN 30%
MORE THAN 25%
8-oz. GLASS BOTTLE
12-oz. ALUMINUM CAN
20-oz. PET BOTTLE
Renew
plantbottle
up to 30% plant-based
100% recyclable bottle
redesigned plastic,
recyclable as ever.
Recover
give it back
TM
Reuse
Each Part is Needed to Make the Chain More Sustainable
LIVE POSITIVELY

Sustainable Packaging Goals

We aim to:

- Improve packaging material efficiency per liter of product by 7 percent by 2020
- Source 25 percent of our PET from recycled or renewable material by 2015
- Recover the equivalent of 100 percent of the containers we produce by 2020
- Achieve 100 percent landfill diversion from our facilities by 2020

LIVE POSITIVELY Coca-Cola

The Next Wave of PET Innovation - PlantBottle®

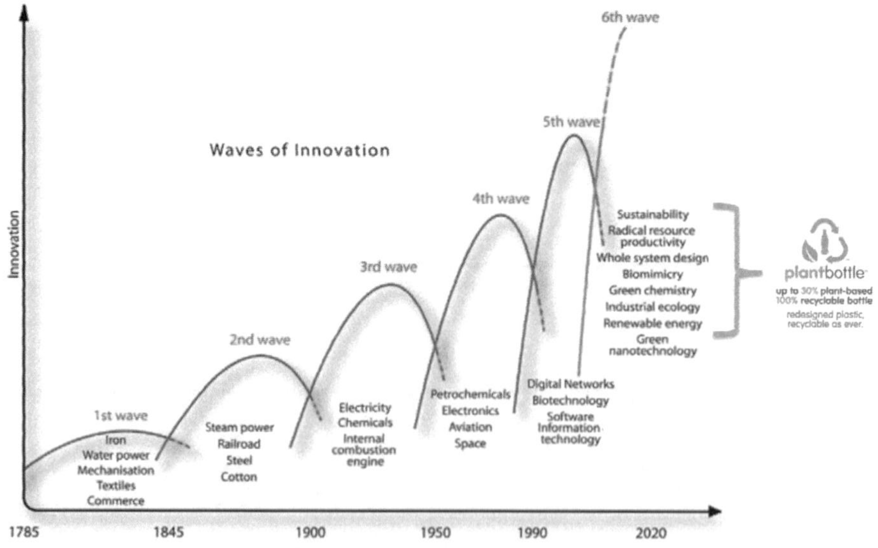

Source: As Time Goes By: From the Industrial Revolutions to the Information Revolution, Chris Freeman and Francisco Louca

LIVE POSITIVELY Coca-Cola

PlantBottle® Packaging

- DASANI PlantBottle made from up to 30 percent plant-based materials nationwide
- Odwalla PlantBottle made from up to 100 percent plant-based materials
- Both 100 percent recyclable - by design!
- ~6.7 billion introduced in 2011, ~160k barrels of oil saved
- By 2020, we aim to transition all of our packaging to PlantBottle.

LIVE POSITIVELY

PlantBottle® - Supply Chain Considerations

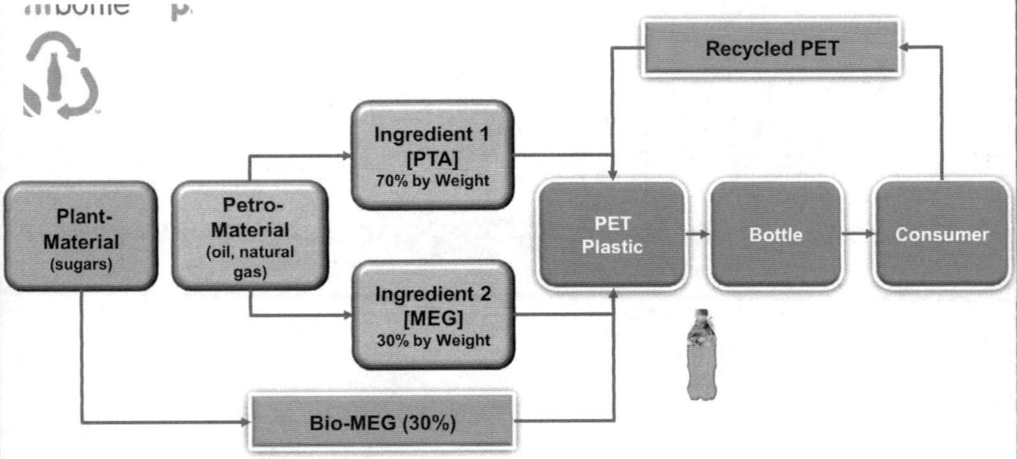

- Key Impact on the overall supply chain is in the resin sourcing and bottle manufacture
- PlantBottle was designed from the beginning to be compatible with the existing recycling infrastructure

LIVE POSITIVELY

Recovery of All of Our Packaging is Critical

Our Performance

- Recover about 38 percent of equivalent bottles and cans we place in market

Our Multi-Faceted Approach

- Established Coca-Cola Recycling
- Encourage recycling through Give It Back
- Educate with Recycling Education Vehicles in 250 cities
- Encourage curbside recycling through:
  - Recyclebank
  - Recycle and Win
  - Bin Grant Program

Coca-Cola RECYCLING

LIVE POSITIVELY Coca-Cola

Thank You!

LIVE POSITIVELY Coca-Cola

Marketplace	Community	Workplace	Environment
Product Integrity & Beverage Benefits			Climate Protection
Active Healthy Living			Sustainable Packaging
			Water Stewardship

LIVE POSITIVELY Coca-Cola

AMERIPEN™
The power of packaging in balance.

The Importance of Packaging To the Sustainable Product Supply Chain

Sustainable Packaging Symposium 2012

J. Pierce
Executive Director, AMERIPEN

AMERIPEN

American Institute for Packaging and the Environment

Represents the Power of Packaging in Balance

- Providing scientific & fact-based DATA
- Taking ACTION that
  - Increases Recovery of Packaging
  - Reinforces the Value of Packaging
- Partners and Collaborates with Existing Organizations

Members

Amcor
Anheuser-Busch LLC
Ball Corporation
Colgate-Palmolive Company*
ConAgra Foods*
DaniMer Scientific LLC
DuPont Packaging & Industrial Polymers*
ExxonMobil Chemical
General Mills Inc
The Hershey Company
Kellogg Company*
McDonald's
Mead Westvaco*
Owens Illinois, Inc
PaperWorks Industries
PARC Corporation
PepsiCo
Procter & Gamble*
Saint-Gobain Containers, Inc
Sealed Air Corporation*
Sonoco
Tetra Pak, Inc*
The Coca-Cola Company*
The Dow Chemical Company*
Waste Management, Inc

* Founding member

AMERIPEN
The power of packaging in balance.

The Supply Chain

AMERIPEN
The power of packaging in balance.

The Hero of the Supply Chain

Packaging as:

- Enabler
- Facilitator

Packaging is part of the SOLUTION

AMERIPEN
The power of packaging in balance.

5

A Sustainable Product Supply Chain

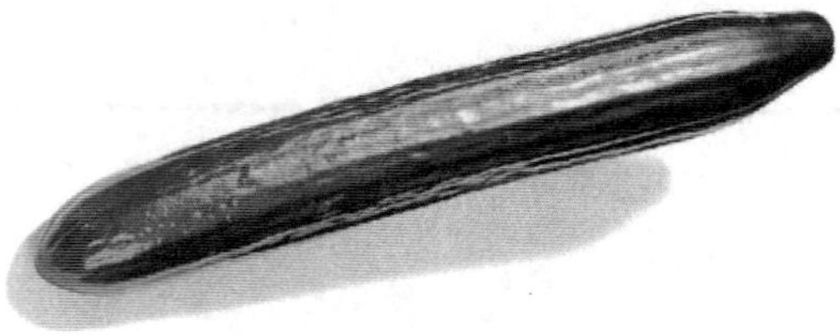

Eliminates Waste

- Product
- Package
- Financial

.......Creates Value

AMERIPEN
The power of packaging in balance.

6

Create Value

- Packaging is playing a greater role in preventing theft *
  - More than $13 billion worth of goods are stolen from retailers each year
  - There are approximately 27 million shoplifters in our nation today

- Packaging is the front line of defense against counterfeit products

* "Information and statistics provided by the National Association for Shoplifting Preventions (NASP), a non-profit organization; www.shopliftingprevention.org."

AMERIPEN
The power of packaging in balance.

7

Value to our Economy

- 750,000+ are employed in the packaging industry in the United States*

- Packaging represents $200 billion or about 2% of the U.S. GDP**

* U.S. Census Bureau
** U.S. Department of Commerce

AMERIPEN
The power of packaging in balance.

8

It's About Optimization, NOT Minimization

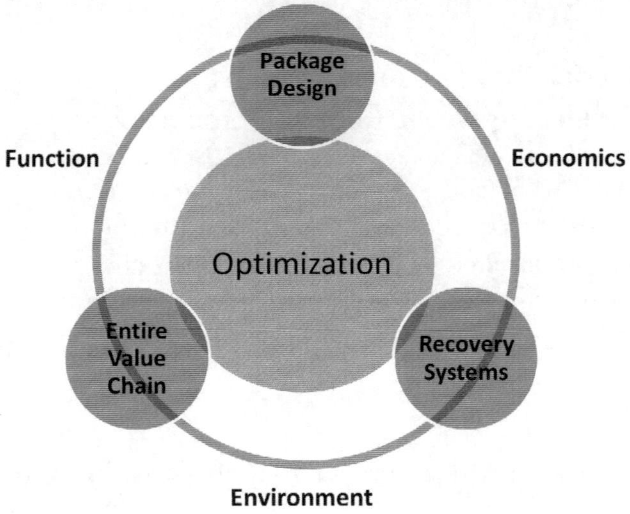

AMERIPEN
The power of packaging in balance.

9

The Power of Packaging in Balance

AMERIPEN
The power of packaging in balance.

10

Optimize Design

- Responsible design is a balance between the amount of packaging used and protection of the product
- Increase packaging to reduce waste
  - If shelf life is improved to reduce product waste
    - 33-50% of food purchased in the U.S. is wasted*
    - $1 trillion of food and beverages are purchased annually
  - If container closure seal integrity is improved
  - If product damage (0.5% – 0.75%) is reduced
- Reduce packaging when performance requirements are not compromised
- Design for End of Life as part of the total lifecycle

* USDA and University of Arizona Garbage Project

AMERIPEN
The power of packaging in balance.

11

Optimize Design – Unsaleables

"The big challenge ...focus on total supply chain process improvement...Five reasons for the current flow (through reverse distribution system) are evident:

- **Products become damaged**
- **Product expiration dates are reached**
- Manufacturers withdraw or discontinue products
- Retailer and wholesaler delist or discontinue products
- Products are recalled"

FMI & GMA, Reverse Supply Chain Improvement, July 2011

AMERIPEN
The power of packaging in balance.

12

Optimize Design - Unsaleables

Six wholesalers and retailers reported unsaleables receipts from manufacturers totaling 1.18% of their total cost of goods last year (2009)

Damaged goods and expired products represent the largest share of unsaleables (43.3% and 37.2% respectively)

FMI & GMA, Reverse Supply Chain Improvement, July 2011

AMERIPEN
The power of packaging in balance.

13

Optimize Design - Unsaleables

- 1.21% of gross sales
- ...Trends affecting and changing the industry in the U.S.:
  - Companies are increasing the use of eco-friendly packaging
    - Potential for increased waste, if the eco-friendly packaging cannot sustain handling and more damages result
    - One of four trends provided in this report

2008 Joint Industry Unsaleables Report: The Real Causes and Actionable Solutions, GMA, FMI, Deloitte

AMERIPEN
The power of packaging in balance.

14

Optimize Recovery

- Design for End of Life as part of the total lifecycle
- Recovery improves environmental performance of the packaging value chain
- Creates jobs

AMERIPEN
The power of packaging in balance.

15

Recovery Creates Jobs

Despite the deepest recession in decades, recycling jobs in North Carolina grew by nearly five percent in the last two years

* Recycling Sector Creates Job Growth During a Down Economy SERDC (Keefe Harrison)

The recycling industry in SC announced >$438MM of capital investment in 2010:

- Created 1,130 new jobs with 28 new or existing companies
- 22 percent increase over the prior year recycling investment of $354 million

* Recycling Sector Creates Job Growth During a Down Economy SERDC (Keefe Harrison)

AMERIPEN
The power of packaging in balance.

16

Optimize Recovery

Complex system

- 24,000 local governments
- Differing governance, management and operational structures
- Differing population densities
- Millions of points of material generation not covered by municipal or private sector services
- Variable levels of technological maturity
- Disaggregated funding and policy frameworks
- Lack of alignment among value chain stakeholders

AMERIPEN
The power of packaging in balance.

17

Optimize Entire Value Chain

- Packaging alternatives must consider the entire value chain
  - Changing one factor in a positive direction may adversely affect performance in another
  - There is no one right material for an application

- A holistic evaluation, assessing the entire system, provides an indication for opportunities to positively affect environmental performance characteristics

AMERIPEN
The power of packaging in balance.

18

Where do we start?

AMERIPEN
The power of packaging in balance.

19

Projects - 2011-2012

- Communicating the value of packaging
- Increasing the recovery of packaging
- Assessing financial platforms for end-of-life packaging

AMERIPEN
The power of packaging in balance.

20

Value of Packaging

Increasing the understanding and communicate the value of packaging

- Communicate value of packaging to key governmental and industrial decision makers
- Develop fact-based information, create communication tools, and identify channels to influence
- Utilize existing information first and sponsor research where needed

AMERIPEN
The power of packaging in balance.

21

Increase Recovery of Packaging

Develop robust data system on U.S. packaging recovery

Top 3 opportunities for each:

- Collection
- Processing
- System Performance

Package Recovery Knowledge Map

AMERIPEN
The power of packaging in balance.

22

Increase Recovery of Packaging

Develop optimized recovery system solutions

- Group best practices into a range of Optimized Recovery System Solutions
- Determine best alternatives to drive used packaging recovery
- Highly ranked solutions as strategy programs
- Cost assessment for optimized recovery system

AMERIPEN
The power of packaging in balance.

23

Assessing Financial Platforms

- The AMERIPEN project will evaluate funding systems that support improved packaging end of life scenarios
  - Data-based and non-judgmental
  - Built upon work by the USEPA and SPC
- The studies will explore the different faces of financial platforms to improve packaging recovery
  - Includes Green Dot/EPR programs, covenants, and simple taxes/fees
  - Also planned are container deposit laws (CDL), Advance Disposal Fees (ADF), and alternative programs such as pay as you throw, recycle bank and others
- Deliverables include
  - Documentation of BMPs and lessons learned for various funding systems
  - Assessment chart of all programs against established metrics

AMERIPEN
The power of packaging in balance.

24

A Sustainable Supply Chain is.....

A World without Waste.....

Values the contribution of packaging.....

Does more with less.....

Recovers valuable used packaging materials

AMERIPEN™
The power of packaging in balance.™

25

AMERIPEN™

The power of packaging in balance.™

Thank You

AMERIPEN™
The power of packaging in balance.™

26

Ice River Springs Water Co.

A Sustainable Approach to Water Bottle Packaging

Ryan L'Abbe – VP & General Manager

April 4, 2012

Our Operations & Customers

- Largest private label bottled water company in Canada (3 plants)
- Established and growing operations in the U.S.(6 plants by end of 2012)
- Privately owned company established in 1995
- Proven supplier to North America's largest retailers for over 14 years

Top US Customers

CVS, Walmart, A&P, ALDI, Save a Lot, Kroger, ShopRite, weis

Operational Excellence

- All plants are fully-owned, vertically integrated, cost competitive and sustainable
- High speed, state-of-the-art production lines located close to customer DC's
- In-house manufacturing of all caps and bottles
- Infrastructure in place to handle large volume production and customer requirements

Water Production Facilities

3

Environmental challenges...

- Consumer choices being driven by impact of packaging
- Water bottles have been the "poster child" of sustainable product packaging concerns

Environmental Initiatives Implemented

✓ Minimize packaging (thinner bottles, smaller caps)

✓ Be energy efficient (lighting, geothermal cooling, green chilling)

✓ Minimize transportation

✓ Integrate recycled content into packaging (next level of thought)

4

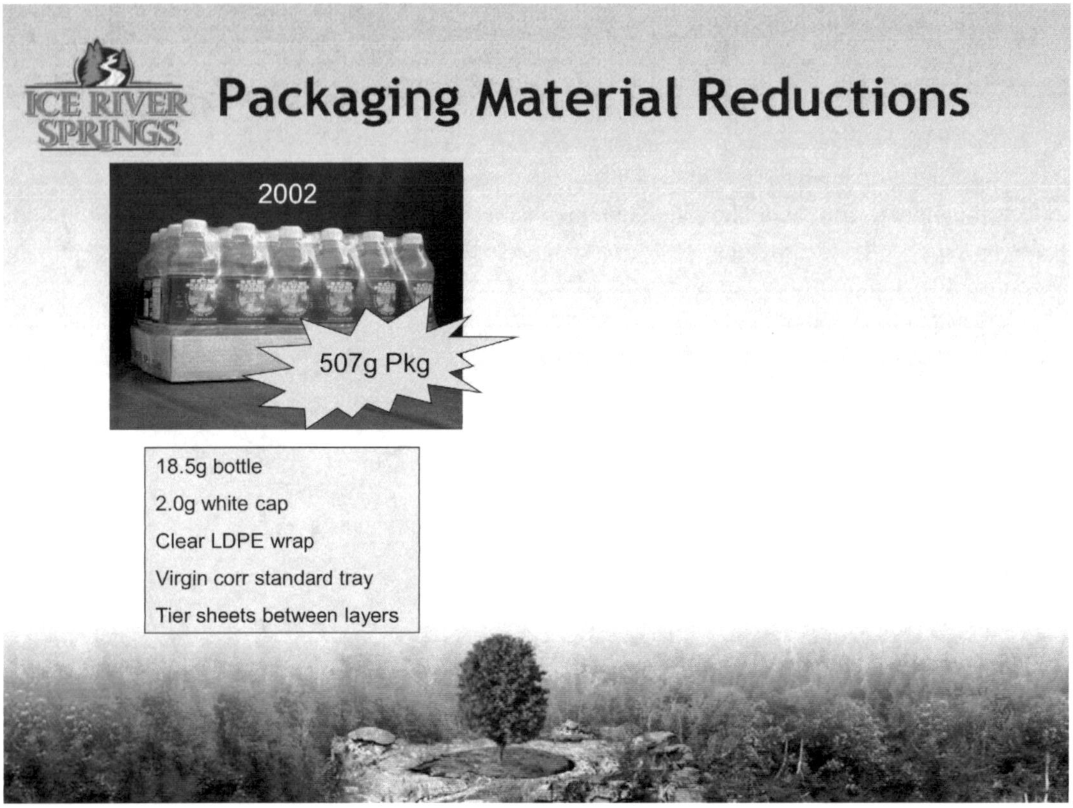
ICE RIVER SPRINGS
Packaging Material Reductions
2002
507g Pkg
18.5g bottle
2.0g white cap
Clear LDPE wrap
Virgin corr standard tray
Tier sheets between layers

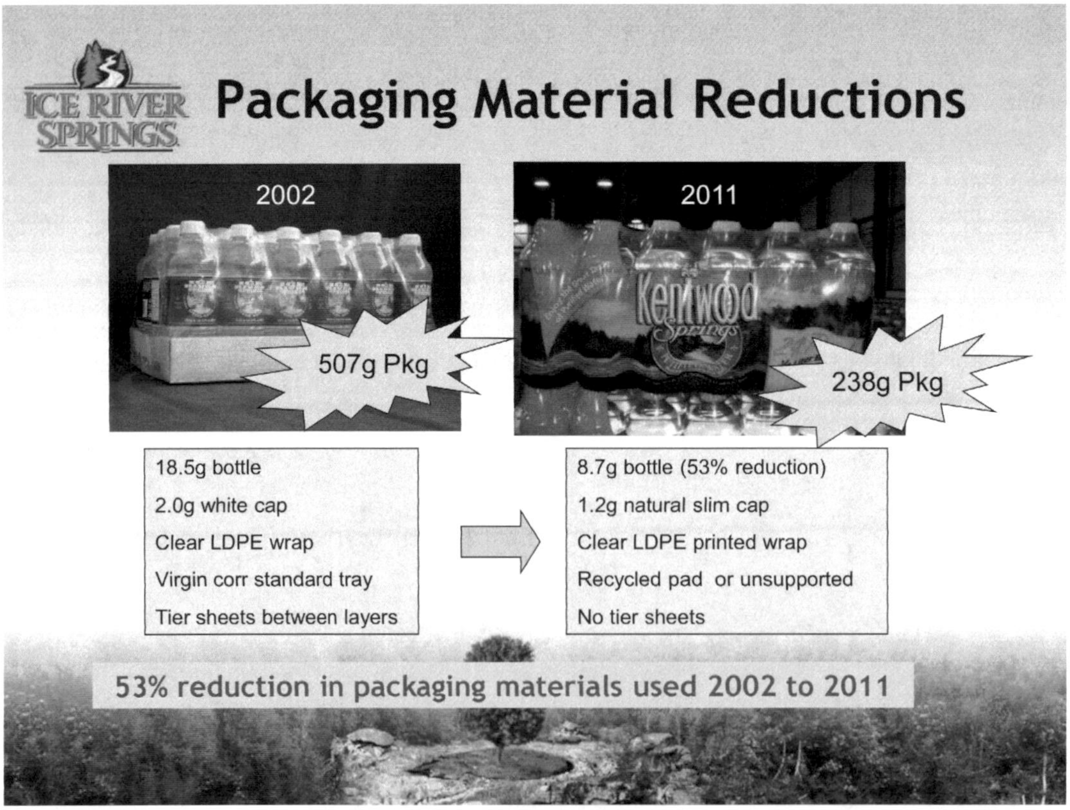
ICE RIVER SPRINGS
Packaging Material Reductions
2002
507g Pkg
2011
Kenwood Springs
238g Pkg
18.5g bottle
2.0g white cap
Clear LDPE wrap
Virgin corr standard tray
Tier sheets between layers
8.7g bottle (53% reduction)
1.2g natural slim cap
Clear LDPE printed wrap
Recycled pad or unsupported
No tier sheets
53% reduction in packaging materials used 2002 to 2011

Integrating recycled content

- Next level of thinking beyond simply reducing the amount of packaging used
- Creates demand for manufacturers of recycled materials
- Creation of truly "green" jobs

How others do it?

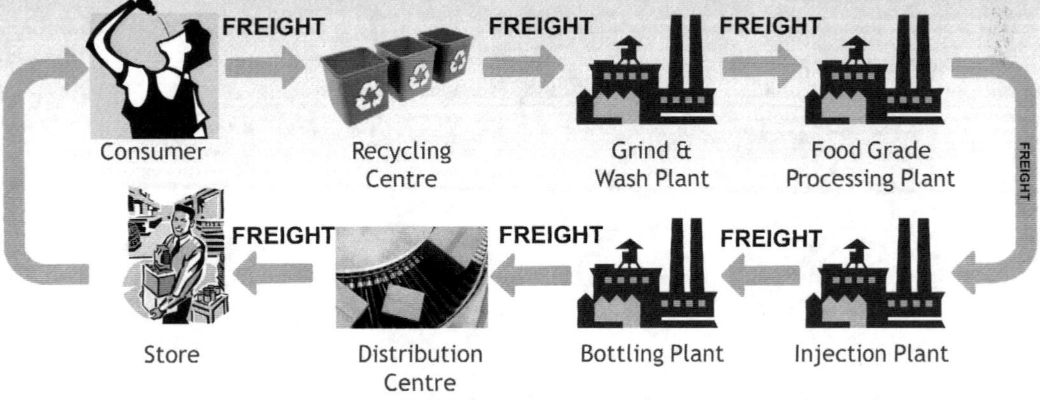

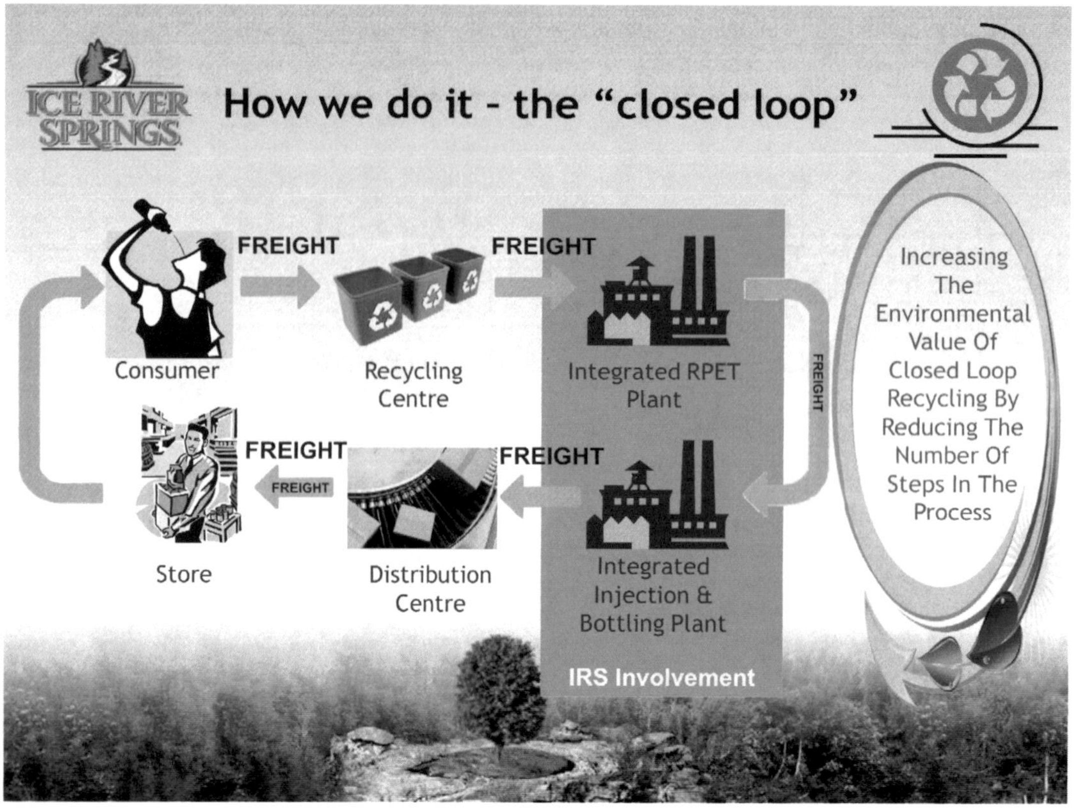
ICE RIVER SPRINGS
How we do it - the "closed loop"
FREIGHT
FREIGHT
Consumer
Recycling Centre
Integrated RPET Plant
FREIGHT
FREIGHT
FREIGHT
FREIGHT
Store
Distribution Centre
Integrated Injection & Bottling Plant
IRS Involvement
Increasing The Environmental Value Of Closed Loop Recycling By Reducing The Number Of Steps In The Process

ICE RIVER SPRINGS
Our "Closed Loop"
WE RECYCLE

Transportation Impact
In-house resin vs purchasing

If you were to produce 1 000 000 cases of 24 pack product from RPET, how many truck kilometres would be required?

Shipping From → To	ICE RIVER SPRINGS NATURAL SPRING WATER	Others using RPET
Material Recovery Facilities → Wash & Grind Plant	386 km	5 640 km
Wash & Grind Plant → Food Grade Processing Plant	0 km (on site)	26 780 km
Food Grade Processing Plant → Injection Plant	5 649 km	2 622 km
Shipping From The Injection Plant → Bottling Plant	2 309 km	19 485 km
TOTAL:	**8 354 km**	**54 527 km**

Vertical integration dramatically reduces the carbon footprint of bottle-to-bottle recycling.

BLUE MOUNTAIN PLASTICS
innovations taking shape

- Subsidiary of Ice River Springs created in July 2010
- Processes incoming bales from municipal recycling plants into clean clear/green flake
- Flake then processed into FDA/CFIA approved pellet for injection at 100% into bottles

Production Facility - Shelburne ON

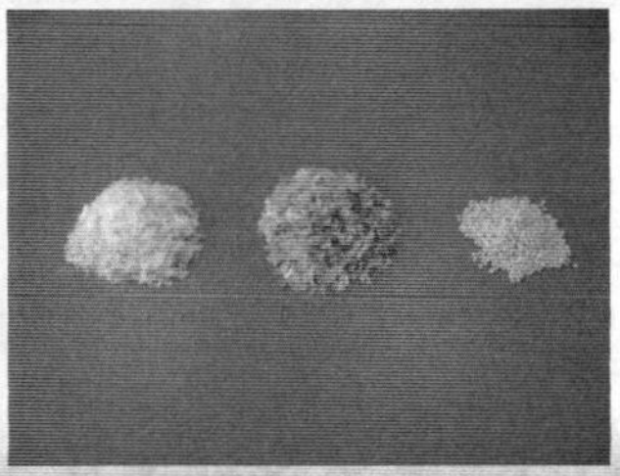

Why have we been successful?

- Sustainability is embedded in our company vision
- Protecting and preserving the environment is imbedded in our culture (owners)
- The leaders of our company place a high internal value on innovation that leads to improvements in sustainability
- Sustainability = reduced costs (long-run)
- Our customers were looking for sustainable packaging options for their brands

Recent Awards

CPIA Canadian Plastics Industry Association

Greener Package 2011 Award Winner

ICE RIVER SPRINGS

Barriers to success…

- Insufficient supply of recycled material from consumers – need to collect more
- Must push to eliminate "look-a-like" plastics that contaminate the PET waste stream (i.e. PVC, PLA, PET-G) – need to drive for PET/HDPE as the plastics of choice
- Customers need to demand reduced packaging
- Transportation costs have a huge impact

J. OTTMAN
CONSULTING

eco innovation
green marketing

20 New Rules in 30 Minutes with Plenty of Time for Questions

Presented to Sustainable Packaging Symposium
By Jacquelyn A. Ottman,
President, J. Ottman Consulting, Inc.
April 4, 2012

Jacquelyn A. Ottman

- Founder and Principal, J. Ottman Consulting, Inc. (1989)
- Author, *The New Rules of Green Marketing* (Greenleaf Publishing, 2011)
- Educator, MCAD, NYDesigns
- Past co-chair, NYC Chapter, o2 Global Network
- Founding Co-chair, Sustainable Brands Conference

J. Ottman Consulting's Clients and Projects

- Kraft, Heineken, Philips, IBM, General Electric
- Toyota, Nike, 3M, Interface, Steelcase, Pitney Bowes
- USDA Certified Biobased, U.S. EPA Energy Star, SmartWay

I. The Green Market

- 83% of U.S. Adults are "Green"
- $290 Billion U.S. Market for Green Consumer Goods—and Growing
- Barriers to green purchasing fast disappearing as quality of green products increases (Coke, Heinz, Wal-Mart)
- But barriers still exist, esp. value perception and skeptism (Greenwash – remember Hefty!).

II. Close the Gaps with a LifeCycle Approach

- Consumers concerned about all life cycle impacts
- Not enough that one attribute is green – Coke Plantbottle is made from renewable materials and is recyclable, too.

III. Communications

- Focus on Primary Benefits; Greener products consistent with better quality of life
  - Tide Coldwater saves money; ZipCar is more convenient; Method package design provides for self-actualization
- Halo brand
  - Odwalla bottle made from sugar cane; Stonyfield Farm yogurt cups help to reduce carbon emissions

IV. Credibility

- Transparency, e.g., Patagonia Footprint Chronicles
- Promote responsible consumption (use and disposal) – Sun Chips composting education; Toyota Prius dashboard, P&G LifeCycle chart/ infographics

ABOUT J. OTTMAN CONSULTING
J. Ottman Consulting, Inc. is a NYC-based marketing consultancy founded in 1989 by Jacquelyn Ottman to help business leaders find opportunity in developing and promoting greener products and lifestyles. Ottman is the author of four books on green marketing including *The New Rules of Green Marketing: Strategies Tools and Inspiration for Sustainable Branding* (Greenleaf Publishing, November 2010), which was recently named by Cambridge University as one of the top sustainability books of the year. The firm has advised over 60 Fortune 500 companies and several U.S. government labeling programs.

HELPFUL RESOURCES AVAILABLE
Want to learn more about green marketing? Visit our website, http://www.greenmarketing.com for articles, blog posts and many helpful links to all things green marketing. Start by downloading our "green paper", "Moving Sustainability Forward" at http://www.greenmarketing.com/articles.

315 East 69 Street, New York, New York 10021 T: 212 879 4160
info@greenmarketing.com http://www.greenmarketing.com

Twitter: @jacquelynottman Facebook: facebook.com/jottmanconsulting

###

###

New Carbon vs. Old Carbon and the term Renewable

Sustainable Products & Solutions Program
Center for Responsible Business

Tony Kingsbury
Dow Executive in Residence,
Haas School of Business, UC Berkeley

Redefining good business.

Old Carbon... or Fossil Carbon

Coal **Natural Gas** **Oil**

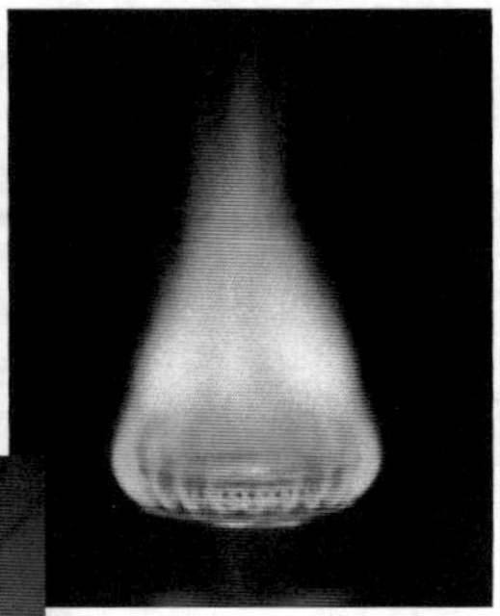

2

SUSTAINABLE PRODUCTS
+ SOLUTIONS PROGRAM
CENTER FOR RESPONSIBLE BUSINESS

Dow

New Carbon… or Bio-based… or Biogenic… or "Renewable"

Trees

Crops

Algae

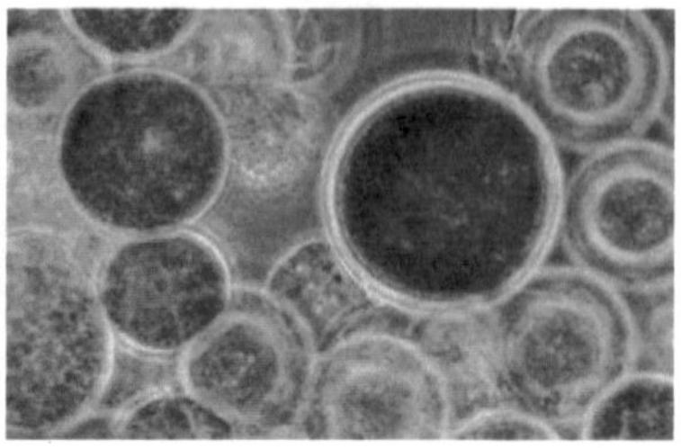

3

SUSTAINABLE PRODUCTS
+ SOLUTIONS PROGRAM
CENTER FOR RESPONSIBLE BUSINESS

Dow

Why do we care about new vs. old carbon?

Three main reasons

1. **We are using up our reserves of "old" carbon in the form of oil, natural gas and coal** … as Thomas Friedman's book *Hot, Flat & Crowded* says… there will soon be 9 Billion people who want the same stuff we have… cars, food, housing, AC, etc. etc. … and all this stuff requires a great deal of "old" carbon resources as produced today
2. **We want to reduce** the carbon footprint and global warming potential (GWP) of the products we produce
3. **Transparency in our supply chains** are beginning to ask for it!

4

SUSTAINABLE PRODUCTS
+ SOLUTIONS PROGRAM
CENTER FOR RESPONSIBLE BUSINESS

DOW

Carbon Transparency Growing

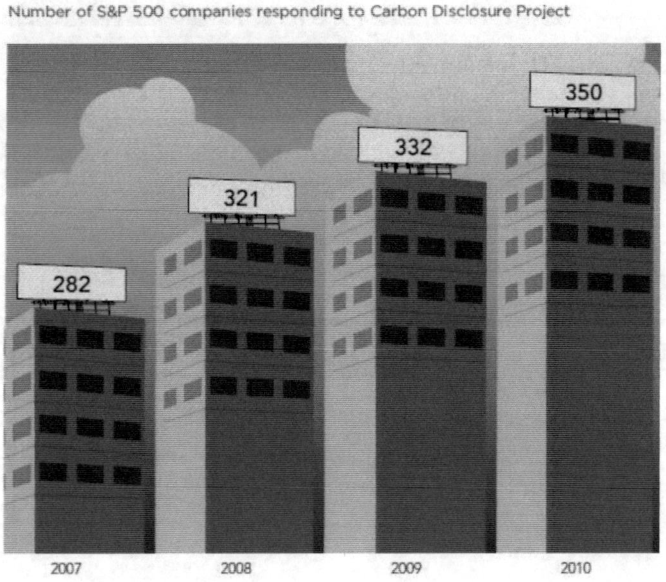

Source: Carbon Disclosure Project & GreenBiz.com

5

SUSTAINABLE PRODUCTS
+ SOLUTIONS PROGRAM
CENTER FOR RESPONSIBLE BUSINESS

DOW

Packaging & New Carbon

- Common Renewable / New Carbon Packages
  - Fiber / Paper based packaging
  - PLA
  - PET ... 30% "New" carbon content in Coke's PlantBottle

Note: This is not the same as buying carbon offsets or RECs

6

SUSTAINABLE PRODUCTS
+ SOLUTIONS PROGRAM
CENTER FOR RESPONSIBLE BUSINESS

DOW

Orange Juice Example

- Oranges are "new" carbon and the paperboard portion of the carton is too…
- Yes …BUT!
- To get that carton to the consumer requires a lot of "old" carbon… fertilizer, distribution transport, electricity, nat. gas, etc.

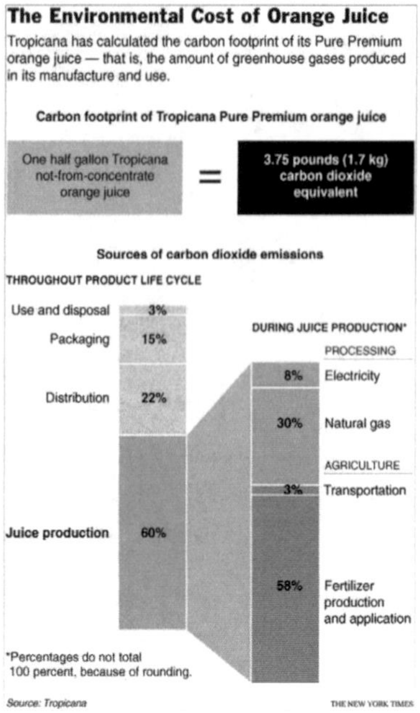

SUSTAINABLE PRODUCTS
+ SOLUTIONS PROGRAM
CENTER FOR RESPONSIBLE BUSINESS

DOW

What is the point?

- Although the primary packaging material may be renewable or "new" carbon… but we need to dig deeper to determine the real global warming potential (GWP):

GWP = Fossil Greenhouse Gas Emissions – Renewable Carbon in the Physical Product

8

SUSTAINABLE PRODUCTS
+ SOLUTIONS PROGRAM
CENTER FOR RESPONSIBLE BUSINESS

16 oz Cup GWP Comparison

- UC Berkeley Analysis of various cups using GaBi software

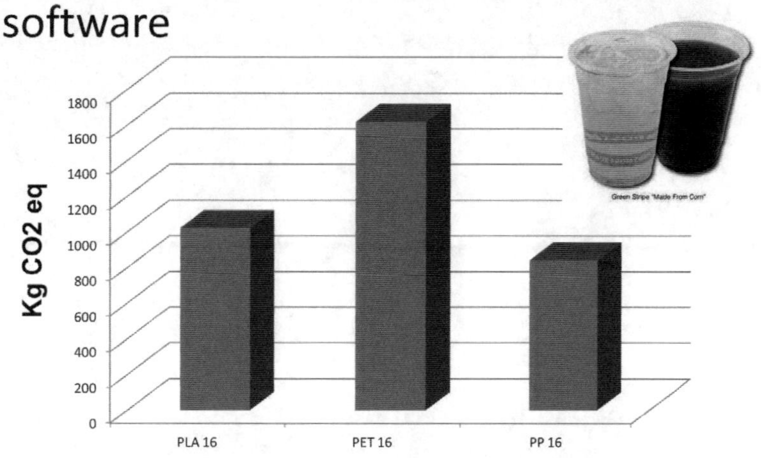

16 oz Cup GWP Comparison

source: LMAS

9

SUSTAINABLE PRODUCTS
+ SOLUTIONS PROGRAM
CENTER FOR RESPONSIBLE BUSINESS

New Carbon Polyethylene from Sugarcane

Dow joint venture in Brazil

- Far less fossil resources than traditional hydrocarbon processes
- Biomass (bagasse) produces heat, electricity and steam for ethanol, ethylene and polyethylene plants

Sustainability Profile

- Produces 1/7th the amount of CO_2
- Most “waste” is used in the process
- Harvesting rainwater & CO2 from air
- Recyclable using existing infrastructure
- Trees adjacent to fields reduce soil erosion
- Sugarcane grown far from rainforests!

10

SUSTAINABLE PRODUCTS
+ SOLUTIONS PROGRAM
CENTER FOR RESPONSIBLE BUSINESS

DOW

What does this mean for packaging?

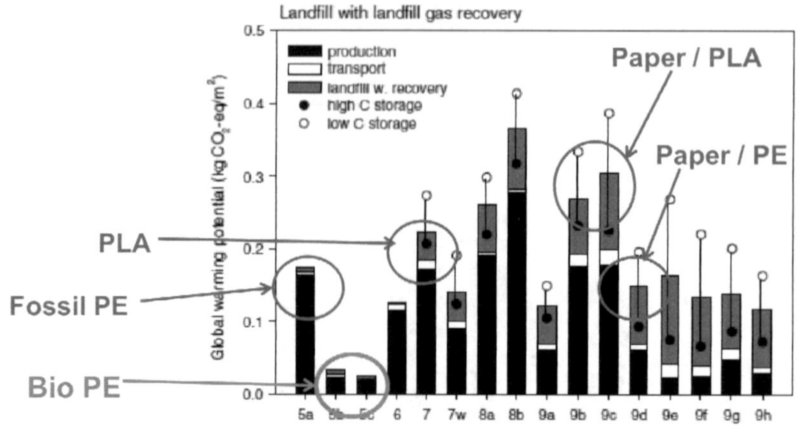

5a	PE
5b	Bio-based PE
5c	Bio-based PE
6	OPP
7*	PLA
8a*	Cellulose
8b*	Cellulose
9a	Paper / OPP
9b*	Paper / PLA
9c*	Paper / PLA
9d	Paper / PE
9e*	Paper / BBP[a]
9f*	Paper / BBP
9g*	Paper / BBP
9h	Paper / EVA

... you need to dig deeper to find out the real GWP of your package

Source: International Journal of Lifecycle Assessments March 2010
Title: Twisting biomaterials around your little finger: environmental impacts of bio-based wrappings
Authors: Hermann, Blok Patel

11

SUSTAINABLE PRODUCTS
+ SOLUTIONS PROGRAM
CENTER FOR RESPONSIBLE BUSINESS

DOW

Don't be Confused

Bio-Based / Renewable Based Packaging are not necessarily biodegradable

- Especially with the rapid expansion of traditional polymers like Polyethylene and PET
- Look for compostable plastics to remain more of a niche market in the coming years due to higher costs and lack of composting infrastructure

12

SUSTAINABLE PRODUCTS
+ SOLUTIONS PROGRAM
CENTER FOR RESPONSIBLE BUSINESS

DOW

Renewable Chopsticks: Scale is important?

✓ Good: Renewable Resource

✓ Bad: Disposable & Growth in Use

Data:

- 45 *billion* disposable chopstick pairs made/year in China, or about 130 million a day,
- 100 acres of trees needed every 24 hours. Think 100 football fields. That works out to 16 - 25 million trees/yr
- The Bring Your Own Chopsticks movement has been gaining momentum, but is still far from succeeding…

Source: Greenpeace & LA Times Article

13

SUSTAINABLE PRODUCTS
+ SOLUTIONS PROGRAM
CENTER FOR RESPONSIBLE BUSINESS

DOW

This is Tough Stuff

"the environmental advantages of bio-based materials are sometimes lower than one may expect because of the comparatively high energy consumption in one or more of the production stages"

2010 Utrecht Univ. paper on biomaterials by Hermann, Blok, Patel

14

SUSTAINABLE PRODUCTS + SOLUTIONS PROGRAM
CENTER FOR RESPONSIBLE BUSINESS

DOW

Conclusions

- Renewable / "new" carbon packaging will grow… but is not a panacea… land & water use, hidden 'old' carbon in supply chains, etc.
- Life cycle understanding of the "old" vs "new" carbon in our products will increasing become important and drive decision making.
- Scale will be a limiting factor for renewables.
- Thus the efficient / functional package, regardless of the material, will usually be the winner in today's business environment.

15

DOW

SUSTAINABLE PRODUCTS + SOLUTIONS PROGRAM
CENTER FOR RESPONSIBLE BUSINESS

University of California Berkeley Haas School of Business

THANK YOU!

Tony Kingsbury
Dow Executive in Residence,
Haas School of Business, UC Berkeley
tkingsbury@haas.berkeley.edu or rakingsbury@dow.com

Redefining good business.

BURT'S BEES®

1 BURT'S BEES®

From our modest beginnings

our commitment to sustainability remains the same

EARLY YEARS

We at Burt's Bees are very committed to the environment and using natural ingredients. Whenever possible, Burt's Bees uses ingredients offered by Mother Nature. We take care of nature, so we can take the best from nature.

2000s

Taking care of the environment is part of our DNA … Natural products work in harmony with our bodies to promote balance and well-being; and we should respect, preserve and improve the precious resource of our environment.

NEW ERA

We believe that nature has the best answers and needs a champion. We are inspired by its huge generosity, its awe-inspiring beauty and strength, and we emulate its resourcefulness. We access and protect it to bring natural solutions to the world.

2 BURT'S BEES®

Brand Leadership

Market Leader	Growth Leader	Thought Leader	Innovation Leader	Consumer Leader
#1 Natural Personal Care brand, nearly double of #2 player	Primary driver of Natural Personal Care growth	Developer in Natural standard, natural vision at retail, and driver of sustainable business practices	Introduced numerous successful, high quality, truly natural innovations over the past 3 years	Extraordinary consumer loyalty and adoration fueling word of mouth and strengthening brand health
#1 Natural brand	**+14%** 5Year CAGR	**ZERO** Waste to Landfill	**75** Items launched over past 3 years	**88%** Brand Awareness
21% Market Share	NPC projected to grow **4X** Faster than non-natural	Carbon Neutral NPA natural seal	Accounts for **1/3** of new NPC item sales since 2008	Ranked **#1** Green Brand

Made of Natural Ingredients

High standard for Natural Ingredients

+95% Natural Ingredients

Source: IRI FDMx CY 2010; Nutrition Business Journal 2010; Millward Brown 2011; 2010 Image Power Green Brands Survey

BURT'S BEES

A New Formula for Growth

Brand & Products

Sustainability

Marketing Stories

UNIQUENESS QUALITY + LEADERSHIP

Consumer Preference & Brand Followers

RESULTS

HEALTH & SAFETY

TRUST

4

BURT'S BEES

Sustainability Key Point of Difference for Brand

"The only **sustainable growth** is consumer-demanded growth. Clearly in a resourcefully strained world we need to think about environmental and social responsibility."

- Keith Weed, CMO Unilever

"Firms with a **clear vision and the execution capabilities** to navigate the [sustainability] megatrend will come out ahead."

- 2010 Harvard Business Review

5

BURT'S BEES

THE NEW, GREATER GOOD MODEL

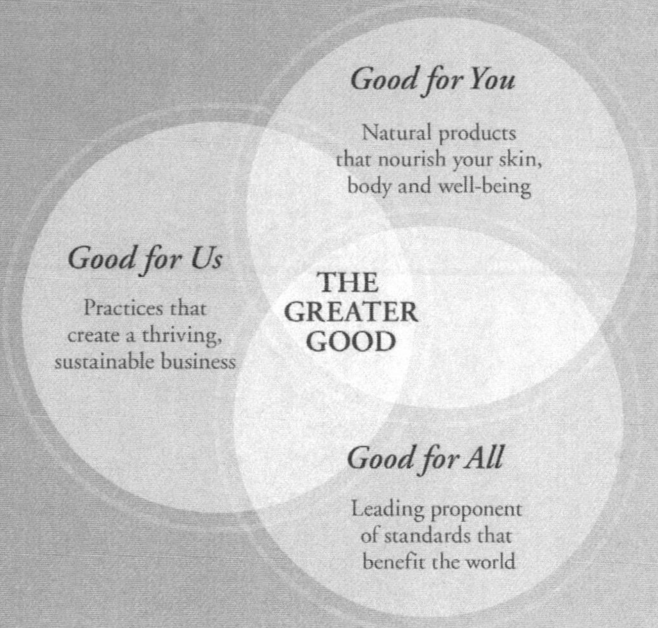

2020 Sustainability Goals

"The ***Greenest*** *Personal Care Company on Earth"*

Engagement	Product
• 100% Employee Engagement	• 100% Natural • 100% PCR/Biodegradable

Carbon	Reduce Waste	Green Building
• 100% Renewable Energy	• Zero Waste	• LEED Certified

BURT'S BEES

FY2012 Environmental Sustainability Goals

ES Dial Score

5% Reduction in Non-Product Water Used ***(5 points)***

5% Reduction in Waste-to-Energy ***(2.5 points)***

>90% Green Derby ***(2.5 points)***

5% Reduction in Total Energy Used ***(5 points)***

BURT'S BEES

8

Recycling Centers

Located in our plants and office areas

BURT'S BEES

Zero Waste to Landfill

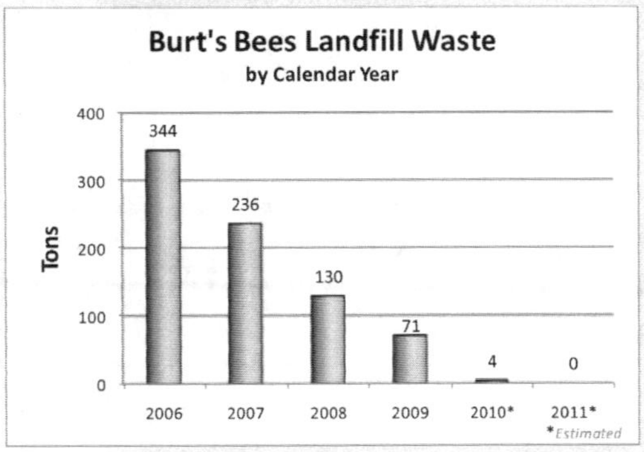

The materials collected in our grey bins are now being shipped to a waste-to-energy facility.

We must recognize that utilizing a waste-to-energy outlet is not our end goal.

Rather this is but one more step in our journey to zero waste.

10

BURT'S BEES

BBI Dumpster Day – April 2008

http://youtu.be/ySfdZh3MbOs

BURT'S BEES

Sustainable Packaging

BURT'S BEES

12

BURT'S BEES

Burt's Bees Packaging Statement

Burt's Bees Packaging Statement:

Burt's Bees is a Global Earth Friendly Natural Personal Care Products Company whose goal is to drive the most Sustainable Packaging guidelines in its packaging choices and decisions.
We strive to use the most Environmentally Responsible/Sustainable packaging systems available that meet or exceed our Quality Standards, as well as Regulatory, Product and Global Customer requirements without sacrificing consumer appeal. We will also use our leadership position to pioneer new systems that will further the development of sustainable alternatives for all to use in the name of The Greater Good. Packaging will not only contain the product but will act as a theft deterrent when possible.

13 BURT'S BEES

Packaging materials that are preferred:

Materials that are highly recyclable, that contain PCR (Post Consumer Recycled) materials, and/or have a low overall environmental impact.

- **Glass: Plentiful resources, uses recycled materials, highly recyclable, and can be reusable.**
- **Aluminum: Plentiful resources, uses recycled materials, highly recyclable and can be reusable.**
- **Polyethylene Terephthalate (PETE #1): Components made from at least 50% recycled materials and that are highly recyclable (bottles with necks).**
- **High Density Polyethylene (HDPE #2): Components made from at least 50% recycled materials and that are highly recyclable (bottles with necks).**
- **Steel: Components made from at least 35% recycled materials and that are highly recyclable.**
- **Paper: Components made from at least 50% recycled materials (40% PCR) and that are highly recyclable.**
- **Degradable Materials: Terra Skin, PHA, PSM (GMO free)**
- **Corrugate: Components made from at least 100% recycled materials (100% PCR)**

Packaging materials to avoid*:

Materials that have limited recycling opportunities, little to no PCR materials are availability, and/or that present a medium to high overall environmental impact.

4 LDPE
- Low Density Polyethylene (LDPE #4): Components made from less than 50% recycled materials with limited recycling opportunities.

5 PP
- Polypropylene (PP #5): Components made from less than 30% recycled materials with limited recycling opportunities.

7 OTHER
- Laminate tubes: laminates made from less than 25% Post Consumer Materials.

*acceptable until replacement materials come into use

BURT'S BEES

Packaging materials that we strive not to use:

Materials that are not made from recycled material and are not recyclable, have very limited or no PCR materials available and/or present a potential environmental impact.

- Polyurethane
- Urea-Formaldehyde
- Polystyrene (PS)
- Polycarbonates (PC)
- Polyvinyl Chloride (PVC)
- Acrylic
- Acrylonitrile Butadiene Styrene (ABS)
- Polyethylene Vinyl Acetate (PEVA)
- Ethylene Vinyl Acetate (EVA)
- Bio-resins made from GMO material
- Virgin board
- Virgin SBS board

BURT'S BEES

Inks & processes that are preferred:

Additives or processes that optimize visual package presentation while minimizing environmental footprint (i.e. reducing weights or number of inks applied, chemical toxicity, energy required to dry inks or coatings, VOC content, etc.)

- Vegetable Based Inks
- Aqueous Coatings

BURT'S BEES

Inks & processes that are acceptable:

Additives or processes that optimize visual package presentation while reducing the environmental footprint (i.e. reducing weights or number of inks applied, chemical toxicity of inks and by-products, energy required to dry inks or coatings, VOC content, etc.)

- Screen Printing (flexible)
- UV Inks
- UV Coatings
- Sustainably Sourced Pearlized Inks
- Sustainably Sourced Pearlized Coatings

BURT'S BEES

Inks & processes that we strive not to use:

Additives or processes with potential to have a negative environmental and/or social impact.

- Vacuum metallization
- Screen printing (on Rigid containers)
- UV metal inks
- Pearlized inks (contains mica)*
- Inks containing GMO materials
- UV coatings
- Solvent based inks
- Phenolic coatings

*If pearlized inks do not contain mica or supplier can provide documentation that mica is sustainably sourced, these may be used

BURT'S BEES

Sustainable Packaging
Packaging Decision Tool

BURT'S BEES

20

BURT'S BEES

Packaging Decision Tool

- A tool has been created to compare like packaging items in terms of cost and sustainability
- For instance, there is a chart for primary packaging types so one can see how a metal closure (i.e. cap or lid) would rate against a plastic closure

BURT'S BEES

Packaging Decision Tool Ratings

- Cost ratings were assigned by the Purchasing team and sustainability ratings were assigned by Sustainability team
- Ratings used for graphs on next slides:

Ratings:		**Cost**
	1	low
	2	moderately low
	3	medium
	4	moderately high
	5	high
		Sustainability
	1	unacceptable
	2	poor
	3	marginal
	4	acceptable
	5	superior

BURT'S BEES

Packaging Decision Tool – Secondary Packaging

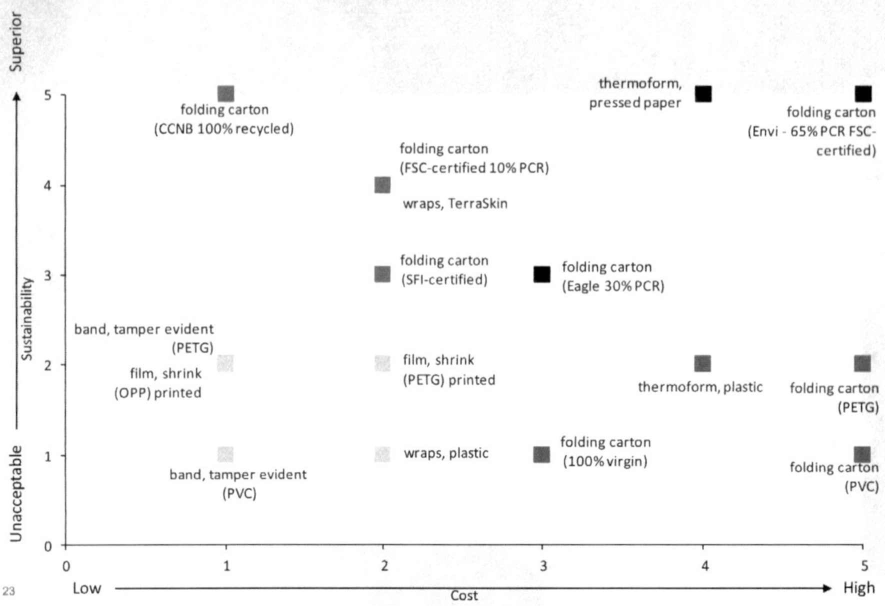

23 BURT'S BEES

Sustainable Packaging Examples

24 BURT'S BEES

Cost Savings & Sustainable Packaging

Completed Projects/Changes	Sustainability I	Sustainability N	Sustainability D	Cost D	Cost N	Cost I
Lip Balm – TE shrink removal	✓			✓		
Lip Shimmer – TE shrink removal	✓			✓		
Lip Gloss – TE cap – shrink removal	✓			✓		
Lip Gloss – TE cap and tube to HDPE #2	✓			✓		
Lip Balm – Common Base		✓		✓		
Lip Shimmer – Common Base		✓		✓		
Lip Shimmer moved to line 800	✓			✓		
Blister Boxes – material change to CCNB	✓			✓		
Blister Boxes – material change 18pt to 16pt	✓			✓		

Ongoing Projects/Changes	Sustainability I	Sustainability N	Sustainability D	Cost D	Cost N	Cost I
Lip Balm – fill reduction 5.0 grams to 4.5 grams	✓			✓		
Lip Balm – New stick supplier			✓	✓		

I = Increase, N = Neutral, D = Decrease BURT'S BEES

Steel Tins – Nibbs

Nibbed Tins

- Average 60% recycled material
- Reduced material from #72 to #65 steel
- Nibb design keeps the lid on the body consistently
- Removal and application forces are more controlled vs. friction fit
- Design allows for shrink film to be removed from domestically shipped products

26 BURT'S BEES

Lip Balm Carton Redesign

Lip Balm Carton Redesign

- Average 10% material reduction
- CCNB 100% recycled material – min 35% PCR
- All carton moved to 16 pt. material
- Case redesign
  - increased pieces per pallet by 4032(1pk) & 7488(2pk)

BURT'S BEES

Gift Kits Redesign

- New Design of our Gift Kits – with improved appearance
- Removed PVC, Phthalate Free PVC or PEVA bags with zippers
- All materials recyclable in the paper recycle stream
- Paperfoam trays
- Cellulose window film
- Easier to assembly – reduced labor costs
- Removed dennison tag seals

BURT'S BEES

Packaging System Improvements

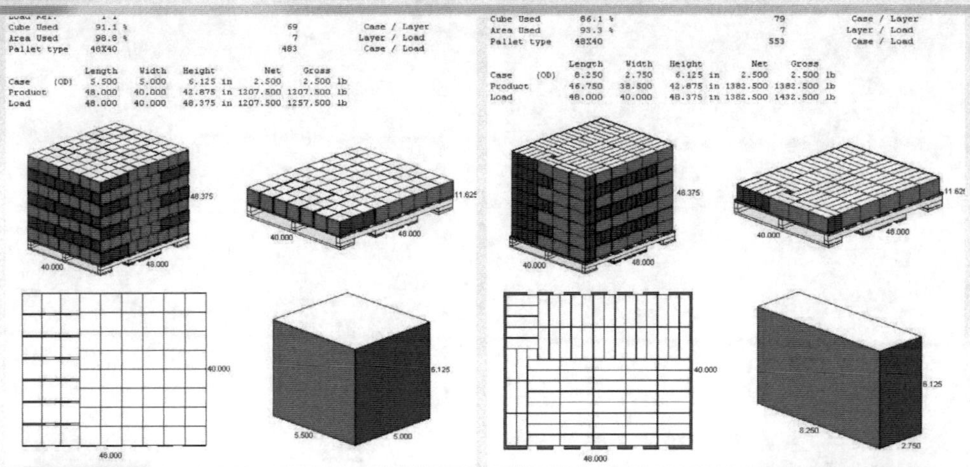

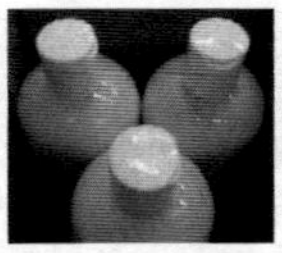

- Shipping Configurations – simple changes can equal transportation and inventory savings
- Design change = 210 more bottles per pallet

BURT'S BEES

Sustainable Packaging Learning's

Virgin 80%PCR 100% PCR

Post Consumer Recycled Materials

- Bottles made from high levels of tend to be "grey".
- Intensity of grey increases with level of PCR.
- Processing is more sensitive
- Color Matching – Must be done in the resin at the final PCR %

BURT'S BEES

BURT'S BEES®

32

BURT'S BEES®

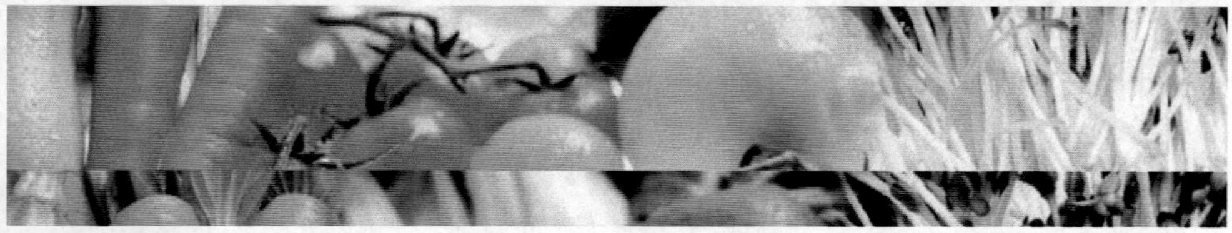

Feeding the Future: The Food Waste Dilemma

Michael Hewett
Director of Environmental and Sustainability
Publix Super Markets, Inc.

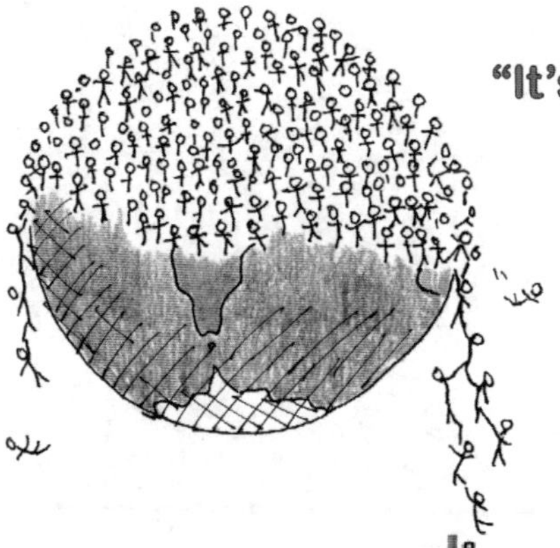

"It's not space we need, it's balance."
National Geographic Magazine

Food Insecurity exists whenever the availability of nutritionally adequate and safe foods or the ability to acquire acceptable foods in socially acceptable ways is limited or uncertain.

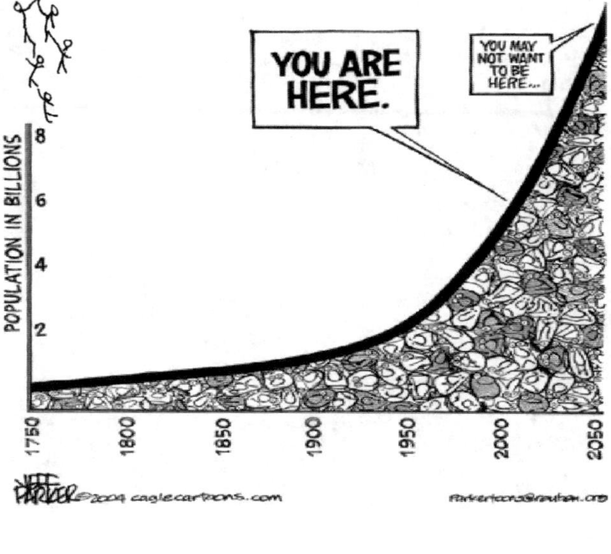

Post Harvest
Food Waste
In the
United States
Disposal = 79.4 billion lbs
Diversion = 42 billion lbs
Total Generation = 121.6 billion lbs

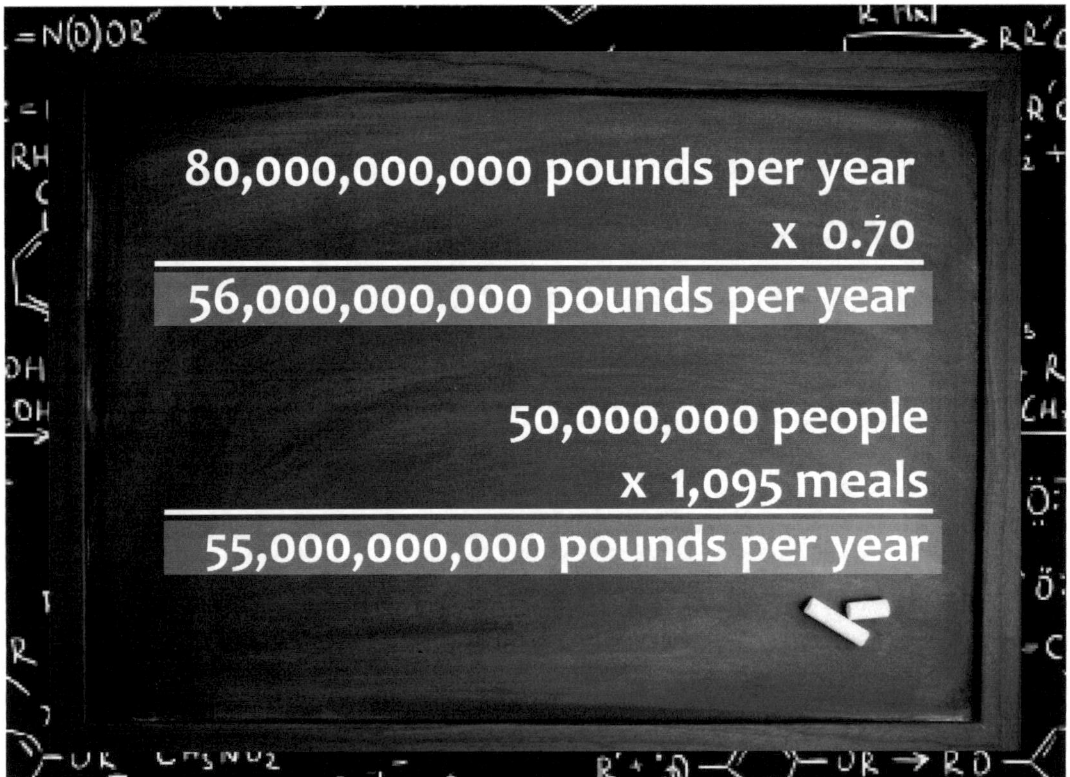
80,000,000,000 pounds per year
x 0.70
56,000,000,000 pounds per year
50,000,000 people
x 1,095 meals
55,000,000,000 pounds per year

…we're literally throwing money away

SOCIETY

- Social and economic cost of hunger and food insecurity in the US in 2010 was **$167.5 billion** (+ federal nutrition programs @ **$94 billion/year**) = **> $260 billion**

HOUSEHOLDS

- 25% of food we bring into our homes is tossed – adding approximately **$500 to $2,000 to a household's food bill per year**

BUSINESSES

- Loss in terms of **food that is not sold**
- Cost of waste **removal fees**

5

…we're wasting energy and water

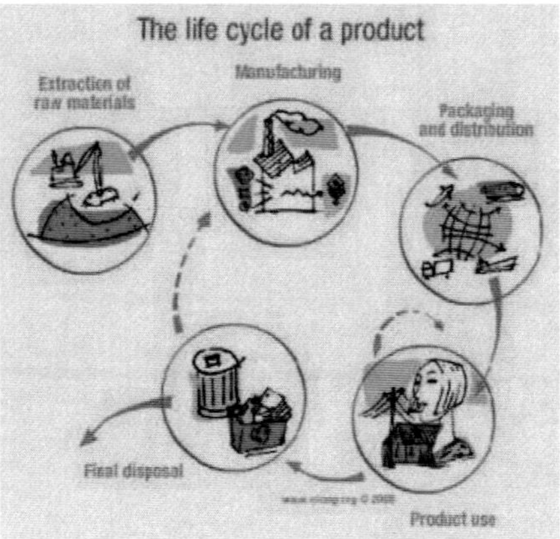

- *Energy* -- Annual U.S. food waste equals energy equivalent of **350 million barrels of oil** (enough to power the whole country for a week.)

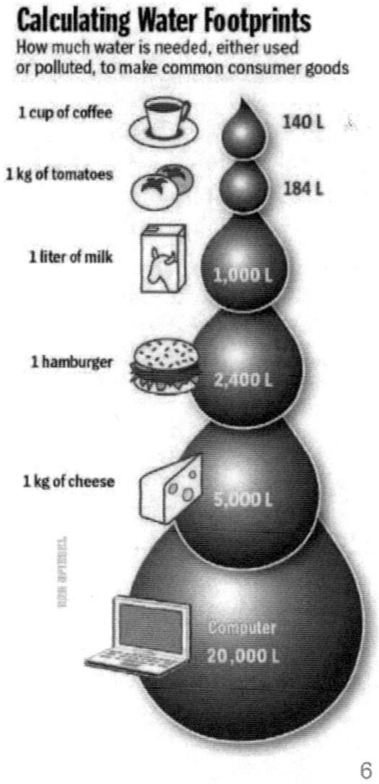

- *Water* -- Estimates are that food waste accounts for more than a **quarter of total freshwater consumption globally.**

6

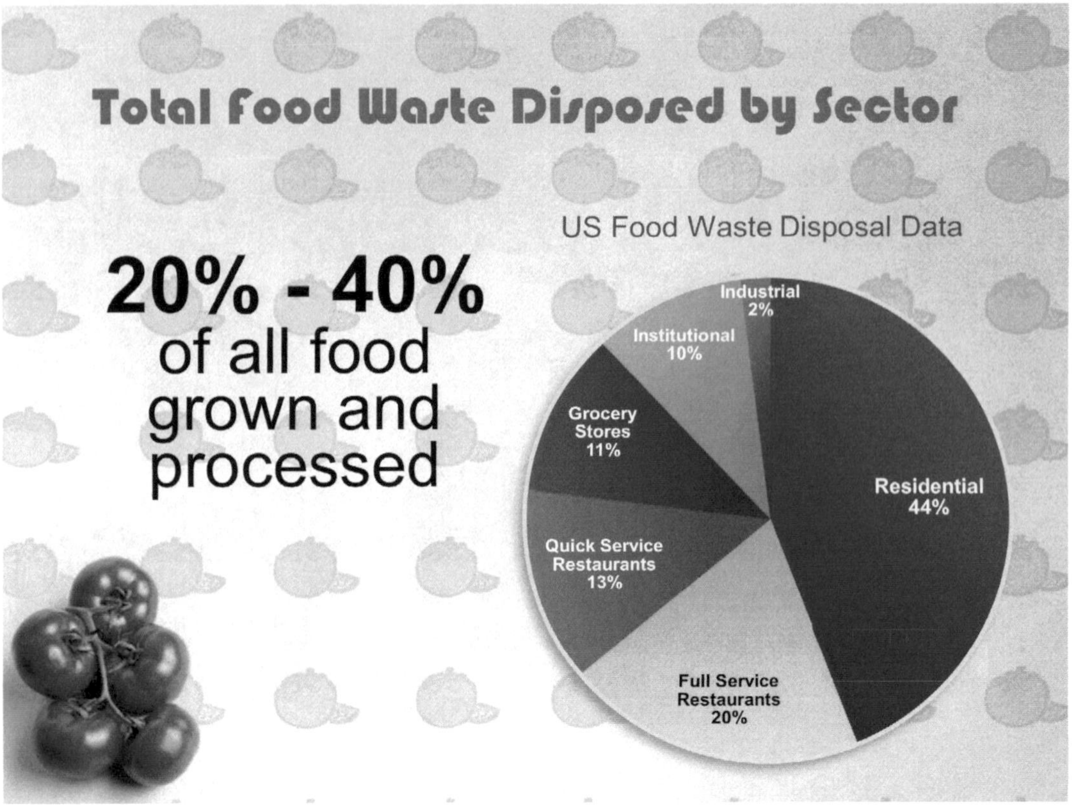

Total Food Waste Disposed by Sector
US Food Waste Disposal Data
20% - 40%
of all food
grown and
processed
Industrial
2%
Institutional
10%
Grocery
Stores
11%
Residential
44%
Quick Service
Restaurants
13%
Full Service
Restaurants
20%

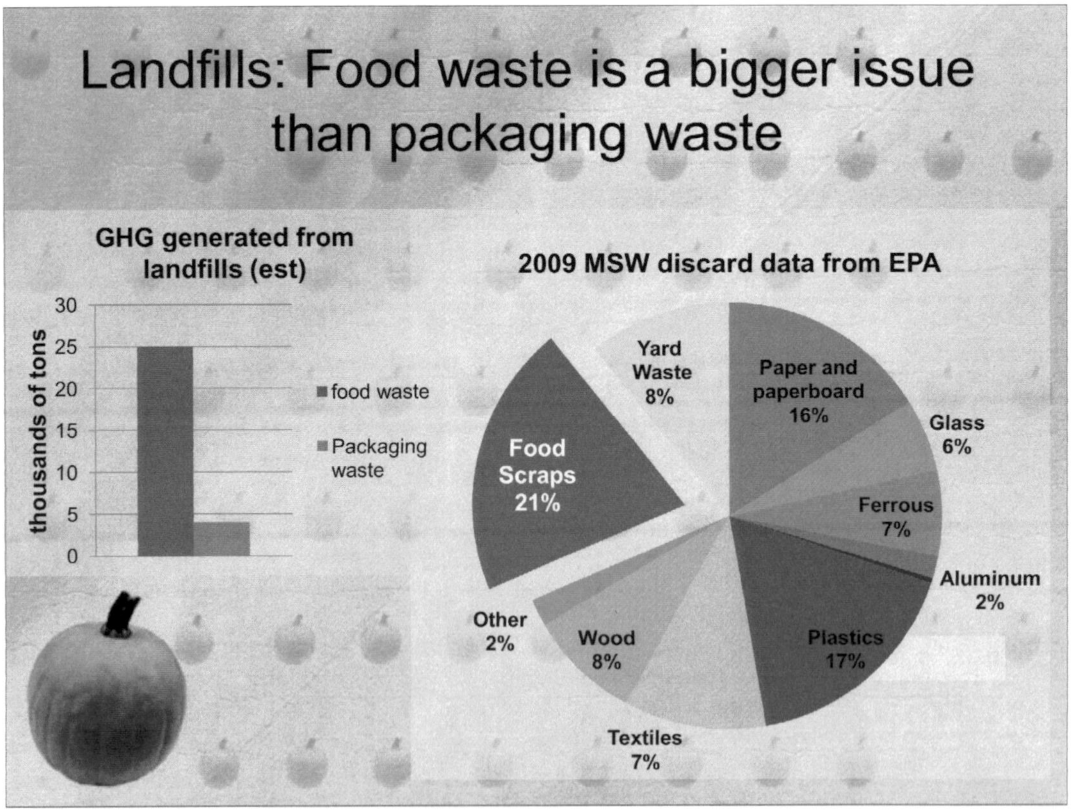

Landfills: Food waste is a bigger issue than packaging waste
GHG generated from
landfills (est)
thousands of tons
30
25
20
15
10
5
0
food waste
Packaging
waste
2009 MSW discard data from EPA
Yard
Waste
8%
Paper and
paperboard
16%
Glass
6%
Food
Scraps
21%
Ferrous
7%
Aluminum
2%
Other
2%
Wood
8%
Plastics
17%
Textiles
7%

The food waste issue is not embedded in the public consciousness yet, but...

<u>Industry Initiatives</u>:

- GMA/FMI/NRA Food Waste Opportunities and Challenges Coalition
- Unilever – "United Against Food Waste"
- General Mills – "Food Waste Awareness Day"

<u>Government Initiatives</u>:

- FAO – "Save Food!" International Congress
- U.S. EPA – "Food Recovery Challenge"

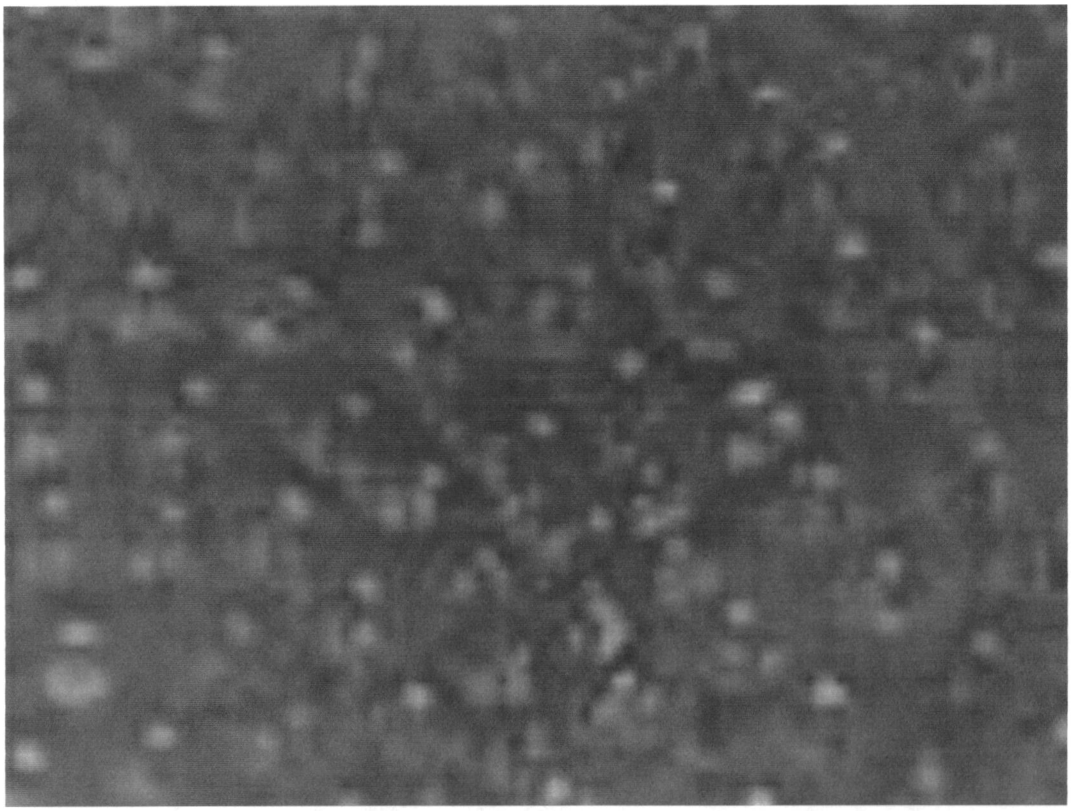

FLORIDA
STRAWBERRY FESTIVAL
Growing Sweet Memories
Mar. 1 - Mar. 11, 2012
PLANT CITY, FLORIDA

2012 Florida Strawberry Festival
Queen and Court

70
70

2012 KING

Queen
Court
Court
Court

U-PICK
PARKING

MARKS &
SPENCER
FOOD
WORLD CUP PARTY
selected party food
buy 3 or more
items & get
購買三件或以上
精選派對美食
15% off

Thank You!
Michael Hewett
Publix Super Markets, Inc.
863-499-8562 (office)
michael.hewett@publix.com

save FOOD

SAVE FOOD 2011 – All stakeholders came together for solutions against food losses.

save FOOD solutions for a world aware of its resources

SAVE FOOD is a joint initiative of the UN Food and Agriculture Organization (FAO) and Messe Düsseldorf GmbH. SAVE FOOD aims at joining forces internationally and discuss the problem of food losses and possible solutions with experts from industry, politics and society both for rich and low-income Regions.

The initiative debuted in 2011 during the interpack show (May 12-18), the world´s leading fair of the packaging and processing industries. It comprised of three elements:

- An **international congress** in Düsseldorf on May 16-17 2011
- An exhibition within interpack incl. the sustainable catering area Food Court
- Two new studies of the FAO on extent, reasons and possible solutions to global food losses (esp. role of packaging in low-income countries)

Top speakers from international organisations and politics generated great media interest.

Modibo Tiémoko Traoré
Assistant Director-General
Agriculture, Food and Agriculture
Organization of the United Nations
(FAO)

Ilse Aigner
Federal Minister of Food,
Agriculture and Consumer
Protection

Prof. Dr. Klaus Töpfer
Former Federal Minister, former
Executive Director UNEP, Executive
Director IASS Potsdam

Johannes Remmel
Minister for Climate Protection,
Environment, Agriculture, Nature
Conservation and Consumer Protection
of the State of North Rhine-Westphalia

Sharaddha Jadhav
Mayor,of Mumbai, India

Global players' executives from the industry presented best practices

Christian Traumann
President, interpack 2011

Philippe Roulet
Head of Global Packaging
Materials & Training, Nestlé S.A.

Nina von Radowitz
Head of Sustainability
and Environment, METRO AG

Friedbert Klefenz
President, Bosch Packaging Technologies

Dirk Löhmer
CEO, COFRESCO
Frischhalteprodukte GmbH & Co.KG

SAVE FOOD becomes FAO's partner in the Global Alliance against Hunger and Malnutrition

- In Fall 2011 the FAO invites SAVE FOOD to join the Alliance Against Hunger and Malnutrition as the only partner from the private sector.
- On October 17 Messe Düsseldorf Chef Werner Dornscheidt addresses the FAO delegates at World Food Day in Rome, the yearly gathering on global hunger themes held on FAO foundation's day.
- Dornscheidt launches the next level of the SAVE FOOD project: from congress to initiative.

The initiative rolls out to the industry and the public

- During FAO's World Food Day SAVE FOOD announces that it will officially become a continous initiative open for subscriptions by corporations and private people to sponsor projects against global food losses.
  The initiative has three goals:
  - Increase knowledge about food loss and waste through sponsorship of studies and regular meetings among all stakeholders
  - Sponsorship of projects in developed and developing countries also together with the partner FAO
  - Organisation of large public events to raise awarness for the issue concurring with the FAO World Food Day. The first one will be held in Düsseldorf in 2012.
- To support the new goals the www.save-food.org webpage is relaunched and social media activities start
- Immediate results: the German Engineering Association VDMA and the packaging corporation MULTIVAC are the first industry supporters from Europe. Still in October the leading US publication Packaging World joins the initiative as a media partner and the American **Food Processing Suppliers Association (FPSA) becomes the first US sponsor.**

interpack – THE NUMBER ONE FOR PROCESSING AND PACKAGING MACHINERY

19 halls devoted to the latest **processing and packaging machinery** and everything about **packaging**

59 countries are represented among the exhibitors – only Düsseldorf offers such a **panoramic international overview**

2.703 exhibitors present their innovations – a source of **new ideas** for the sector.

84 % of visitors are **decision makers** – the ideal prerequisite for doing good business.

97 % of the visitors were **satisfied** in 2011.

interpack
PROCESSES AND PACKAGING

HALL PLAN

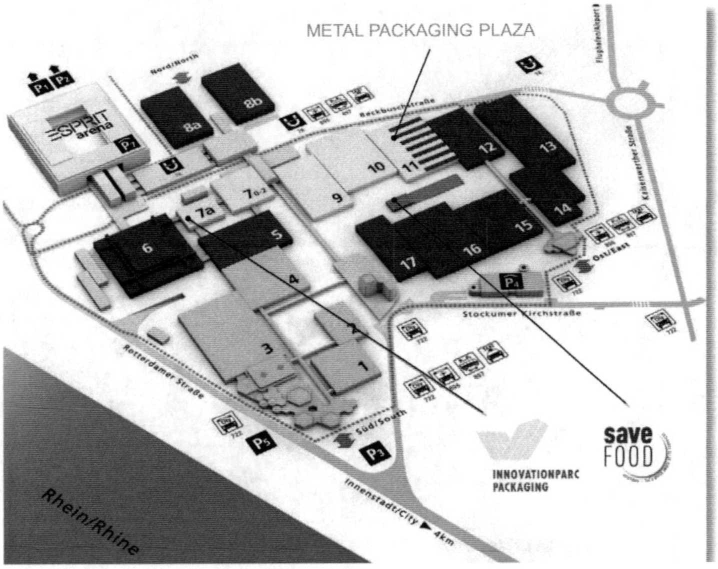

- Processes and machinery for the packaging of: food and beverages, pharma and cosmetics, consumer goods (non-food) and industrial goods
- Packaging materials, packaging means and packaging production
- Processes and machinery for confectionery and bakery goods

Basis for Business
Messe Düsseldorf

interpack
PROCESSES AND PACKAGING

Important dates on the road to interpack 2014

October 2012:
Online application process starts

February 28th, 2013:
Official closing date of applications

June – August 2013:
Placement Information Meetings

October 2013:
Dispatch of admission documents

May 8th – 14th, 2014:
Duration of interpack 2014

Basis for Business
Messe Düsseldorf

AIChE

Institute for Sustainability

Benchmarking Sustainability of Conferences and NGO's

Darlene Schuster-- Executive Director, IFS and CEI
darls@aiche.org
410 458 5870

Industrial Technology Groups (ITGs): Addressing Critical Issues of Today and Tomorrow

Energy

Environment & Sustainability

Water

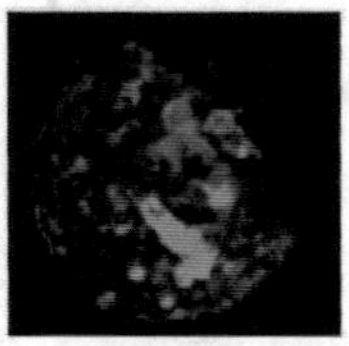

Health & Nutrition

New **Center for Energy Initiatives**

Institute for Sustainability

New **Water Advisory Board**

Society for Biological Engineering

IfS, CEI, Water

- IfS --2005
  - Chair: Deborah Grubbe
  - Launches products to meet needs of Sustainability Professionals
  - Center for Sustainable Technology Practices (CSTP)
  - AIChE Sustainability Index
  - Sustainable Engineering Forum
  - Body of Knowledge and Credentials project

institute for**sustainability**
An AIChE Technological Community

- CEI -- 2010
  - Chair: Dale Keairns
  - Provides an overarching coordination of AIChE energy activities
  - Join an AIChE Division..you are engaged.

- Water Initiative --2011
  - Chair: David Klanecky, Dow
  - Creating a Net Work of Networks

AIChE
Water Initiative

IfS and CEI are also Leaders in Collaboration

institute for**sustainability**
An AIChE Technological Community

Measuring Sustainability Performance Key Questions

- What is the system you want to evaluate?
  - Companies/Organizations
    - *AIChE Sustainability Index Benchmarking*
  - Processes, Development, Products
    - *CSTP Sustainability Roadmap*
    - *ICOSSE Label –ACHEMA 2012*
  - Individuals, Sustainability Practitioners
    - *IfS Body of Knowledge and Certification*
- What are the metrics appropriate to the system under consideration?
    - *Develop of metrics for the above projects*
    - *Biofuels Metrics Roundtable*
    - *2012 Initiative --Sustainability Metrics for Energy Systems*
- How to Validate the metrics?
    - *Peer Reviews*

100's or 1000's of metrics.
What are the 6 -8 Key metrics?
Validity and variance of the data?

Concept: AIChE Sustainability Index^{TM}

Benchmark

- Based on the Wright Killen Refinery Survival Index (Oil and Gas Journal)
- Meant to fill a gap:
  - Relies more heavily on quantitative performance indicators
  - Provides a *heavier weighting* applied to various indicators of safety and environmental performance and to technological innovation towards SD
  - Available to range of sectors, company size
  - The scoring-- robust methodology, designed to account for subjectivity ,

    transparent
- Designed to avoid the 'black box' problem of other indices
- Intended for executives and directors to manage company business lines
- Global, initially focusing on major chemical companies
- Publication remains silent on individual company ratings
- Companies find it useful to benchmark themselves relative to a set of companies

Research funded by United Engineering Foundation

institute for**sustainability**
An AIChE Technological Community

Elements: AIChE Sustainability Index™

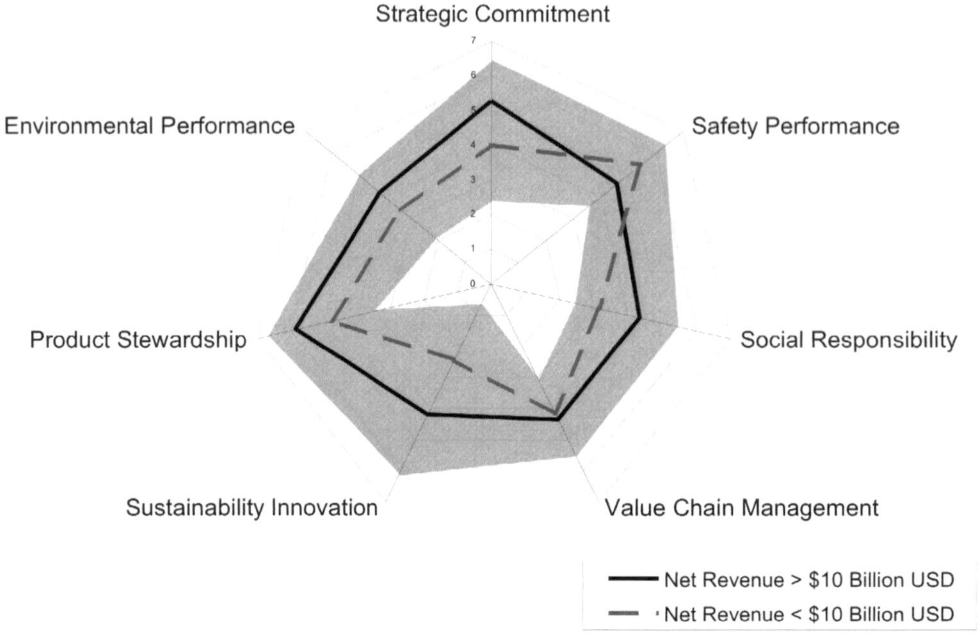

institute for**sustainability**
An AIChE Technological Community

International Certificate for Sustainable Standards for Engineering

Expansion of The AIChE SI from Companies to Products and Processes

In conjunction with DECHEMA

Advisory Board: industry, DECHEMA World Bank USEPA

ICOSSE
International Congress on Sustainability Science & Engineering

Announcing the ICOSSE Label at ICOSSE '12

18 - 22 June 2012
Frankfurt am Main
Germany

Held in conjunction with ACHEMA 2012

International Certificate on Sustainable Standards for Engineering (ICOSSE)
The American Institute of Chemical Engineers (AIChE) and Society for Chemical and Biotechnology (DECHEMA) is launching the "International Certificate on Sustainable Standards for Engineering (ICOSSE)" label at ACHEMA 2012. ACHEMA 2012 is the 30th International Exhibition Congress on Chemical Engineering, Environmental Protection and Biotechnology. The ICOSSE certificate will be offered based on the credible AIChE Sustainability Index which consists of a questionnaire, application form, and evaluation by an advisory board. Designated sustainability products, processes, or services will receive recognition and be displayed on a dedicated website along with the certificate logo.

The purpose of the Congress is to provide a common platform for practitioners of various physical and ecological sciences, engineering fields, economics, and social sciences to exchange emerging ideas about ways and means of protecting the environment and its resources so that humans can achieve sustained economic growth and societal benefits through generations. The focus of this Congress is managing natural resources sustainability from a systems perspective using scientific and engineering innovations. ICOSSE'12 will be offered at ACHEMA with an intensive one day of sustainability programming.

For more information about ICOSSE 2012 go to: http://icosse.org/

institute forsustainability
An AIChE Technological Community

Benchmarking of Conferences and NGO's

Expansion of The AIChE SI from Companies to Organizations

In conjunction with DECHEMA

Advisory Board: Engineers Forum on Sustainability

Key Factors Identification:

- Strategic Commitment
- Conference Performance
  - Contracting
  - Food Waste
  - Energy
  - location and travel
  - Water Usage
  - Paper
- Education Activities
- Headquarters Operations
  - Buildings
  - Employee Travel
- Membership
- Networking and Knowledge

Green Seal Certification

- **In Response To Increased Number Of RFP's The Hilton Americas – Houston Developed A Green Committee in Q4 2007.**
- **The Goal: To Become The First Green Seal Property In Texas**
- **Green Seal: Is an independent non profit organization dedicated to safeguarding the environment and transforming the marketplace, promotes the manufacture, purchase and use of environmentally responsible products and services.**
- **To Qualify: The hotel must demonstrate sustainable practices in the following areas : waste minimization, reuse & recycling; energy efficiency, conservation & management; management of fresh water resources; waste water management; hazardous substances; and environmentally sensitive purchasing.**

Green Seal Certification

Hilton Americas – Houston Underwent The Following:

- **Recycling plastic, paper, all metals and glass**
- Composting food waste
- **Replacing current guest room lighting with energy efficient, Compact Florescent Lighting**
- **Recycling florescent bulbs**
- **Reusing copy paper**
- **Setting ALL copy machines / printers to double sided printing**
- **Purchasing recycled copy paper**
- **Placement of recycle cans in all guest rooms offering our guests the opportunity to recycle.**
- **Installed a gray water reclamation unit in the laundry to reuse waste water form the washers.**
- **Installed a heat exchange which reuses hot air form the dryers to reheat them using less energy.**
- **Water conservation includes using 2.2 gpm faucets, 1.6 gfl toilets and 2.5 gpm showerheads**
- **Linen & towel re use program**

Green Seal Certification

- **Hilton Americas – Houston Underwent The Following:**
  - **Using nontoxic and biodegradable cleaning products**
  - **Following an environmental purchasing policy**
  - **Using environmentally responsible suppliers**
  - **Using materials such as menus, key cards, bath tissue and facial tissue made from post – consumer recycled content.**
  - Replacing disposable food service items with biodegradable items
  - **Installing on/off timers for lighting in low traffic areas**
  - **Recycling liquid shortening**
  - **Recycling toner cartridges**
  - **Recycling batteries**
  - **Serving water only on request in restaurants**
  - **Donating partially used amenities to local shelters**
  - **Replacing equipment with Energy Star rated models**
  - Donating left over food to local shelters at client request
  - **Motion sensor in rooms shut off AC if there is no movement detected in the room**

- ***Since inception The Hilton Americas – Houston has recycled 579,760 lbs or 290 tons of cardboard and have sent over 17,199 - 96 gallon toters or 645 tons to be recycled of mixed plastic, paper, metal and glass.***

- ***Since inception of the Food Compost Project, approximately 935 cubic yards have been shipped to Nature's Way – that equals about 619 tons!!***

- ***In the last 24 months we have recycled 46,838 lbs of kitchen oil or a total of 23.4 tons!!***

- ***Since inception of the water reclamation system in Arpil of 2010 the laundry has saved or 18 million gallons of water.***

- ***In April of 2011 we installed a heat reclamation unit is laurdry saving $550K in electricity and $48K in gas.***

Why?

- **Meetings have an social / environmental impact on our communities. As a result groups like MPI and GMIC (The Green Meetings Industry Council) have been educating companies and meeting pros on sustainability practices.**

- **A five day, 2,500 person conference will use 90,000 cans/bottles; 750,000 cups; 87,500 napkins** Source: Meeting Strategies Worldwide

- **It's the right thing to do:**
  1. For today
  2. For tomorrow
  3. For our future generations

Resources

- WWW.MEETGREEN.COM
- WWW.BLUEGREENMEETINGS.ORG
- WWW.USGBC.ORG (US GREEN BUILDING COUNCIL)
- WWW.EPA.GOV/OPPT/GREENMEETINGS.COM
- WWW.GREENMEETINGS.INFO

Sustainability: An emerging trend?

All the benefits & blessings flowing from the use of the earth were held to be the rightful heritage of all generations

The Old Testament- Genesis & Deuteronomy

"...treat the earth as though we are tenants, rather than owners"....we must leave behind 'enough and as good' for others"

John Locke, Two Treatises of Government, 1689

"Then I say the earth belongs to each...generation during its course...no generation may contract debts greater than may be paid during the course of its own existence"

Thomas Jefferson, 1789

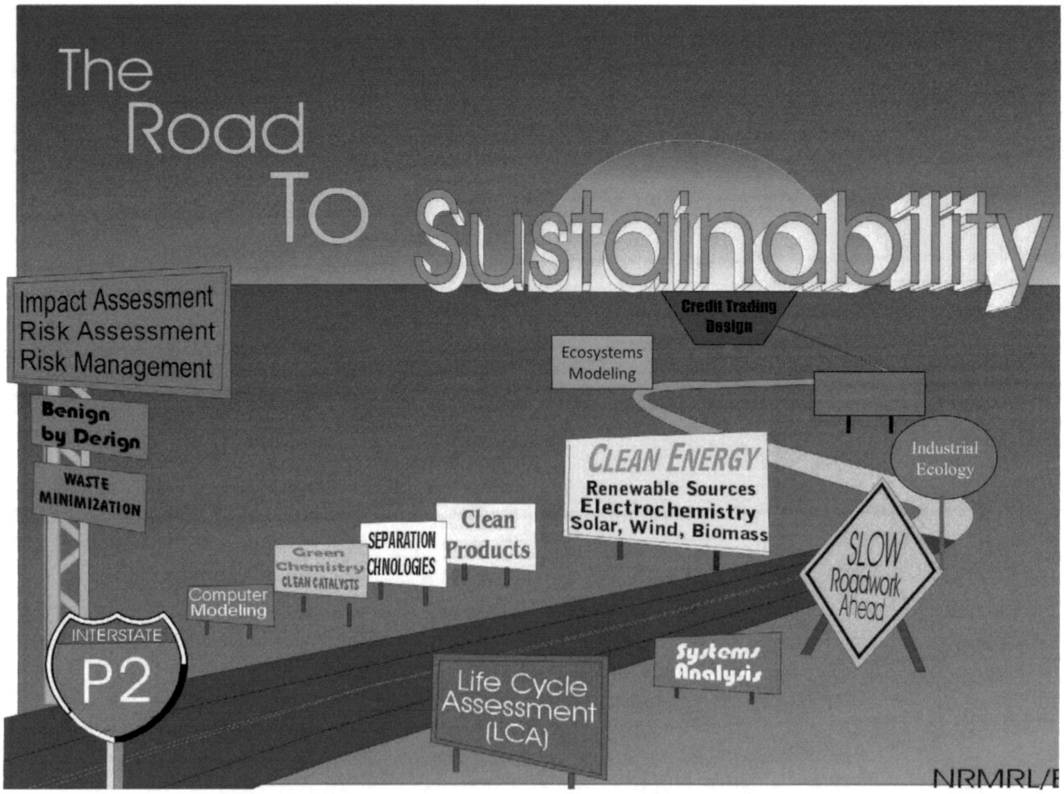

institute
for**sustainability**
An AIChE Technological Community

Overview of the Institute

Mission: to serve the needs of and influence the efforts of professionals, academes, industries, and governmental bodies that contribute to the advancement of sustainability and sustainable development.

SEF	CSTP	YCOSST	Projects of IfS
745 AIChE members 950 Other professionals $25 dues	Industry Group 35 companies 9 roundtables putting sustainability into practice	Partnership w/ SustainUS 9750 students, Incorporates sustainability into undergrad research, sponsors awards & student chapters	1.Sustainability Index 2. ICOSSE Aug 09 3. Founder Societies CM 4. Engineers Forum on Sust. 5. Sustainable Packaging Symposium *6. Credentials and BOK*

Ifs develops, incubates, and leverages sustainability products and services

EARTH911

THE FACTORS THAT AFFECT RECYCLABILITY

Sustainable Packaging Symposium
April 5, 2012
Houston, TX

Earth911
The Recycling Authority

ABOUT EARTH911

THE RECYCLING AUTHORITY

QUICK FACTS

WHO WE ARE

Established in 1991

North America's largest recycling directory:
1 million+ ways to recycle 300+ materials – What, Where, & How

Maintained by a dedicated data team

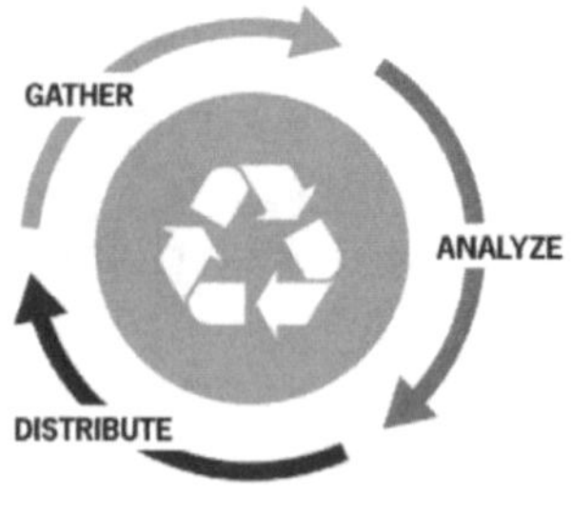

Earth911
The Recycling Authority

Material Recycling Reports
Earth911 Recycling Directory
Recycling Locator
iRecycle® Mobile App
Earth911.com
1-800-CLEANUP
iRecycle
Earth911
The Recycling Authority

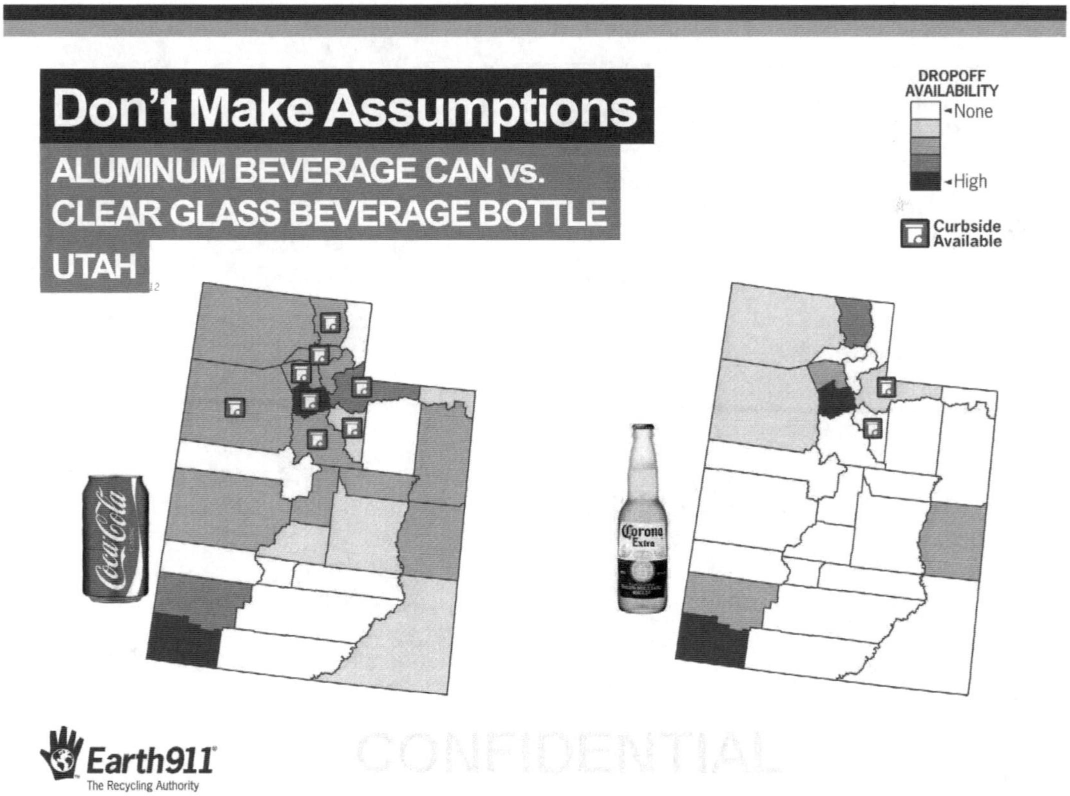
Don't Make Assumptions
ALUMINUM BEVERAGE CAN vs.
CLEAR GLASS BEVERAGE BOTTLE
UTAH
DROPOFF AVAILABILITY
None
High
Curbside Available
Earth911
The Recycling Authority
CONFIDENTIAL

FACTOR 1: RESIN or GRADE

#1 PLASTIC BOTTLE vs. #5 PLASTIC BOTTLE

SOUTH CAROLINA

DROPOFF AVAILABILITY
◂None
◂High
Curbside Available

Earth911
The Recycling Authority

FACTOR 2: ADDITIVES

PAPERBOARD: CEREAL BOX vs. FROZEN FOOD BOX

CALIFORNIA

DROPOFF AVAILABILITY
◂None
◂High
Curbside Available

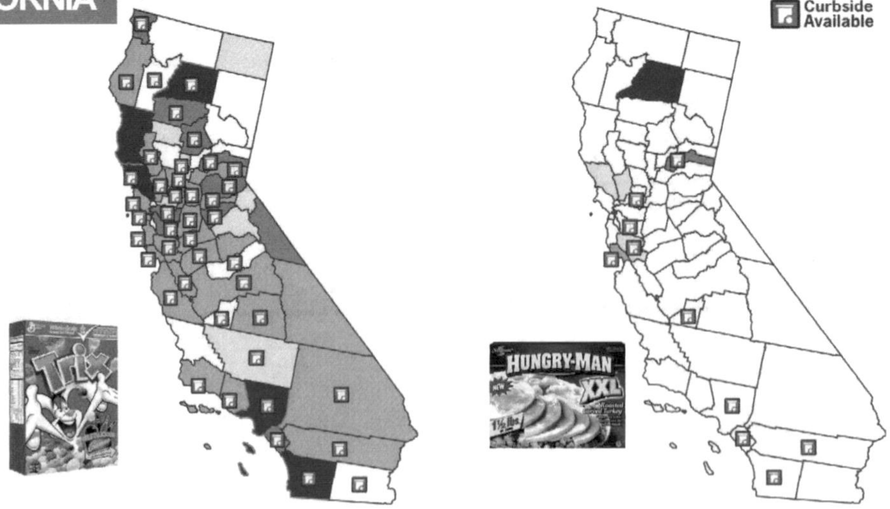

Earth911
The Recycling Authority

FACTOR 3: FORM

HDPE JUGS vs. HDPE TUBS

TEXAS

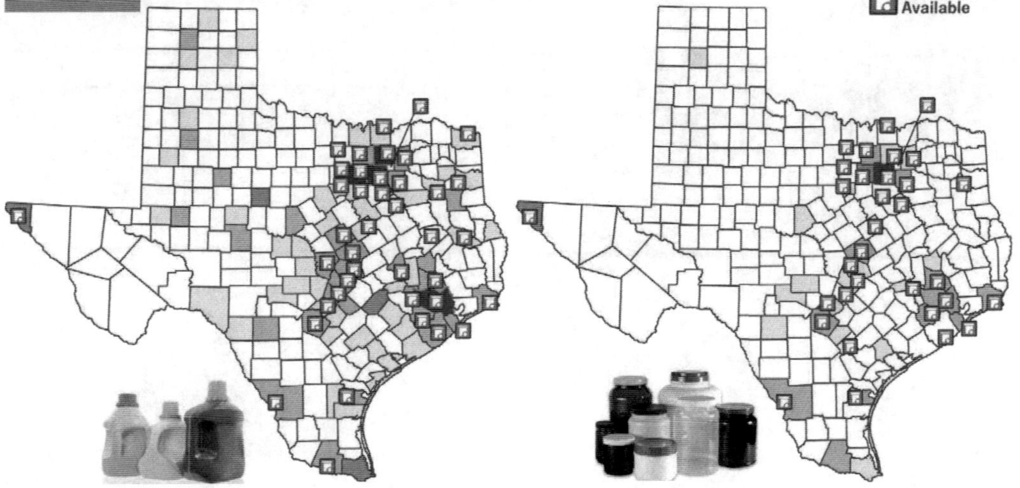

Earth911
The Recycling Authority

FACTOR 4: COLOR

CLEAR GLASS vs. BLUE GLASS

SOUTH CAROLINA

As of March 20, 2012

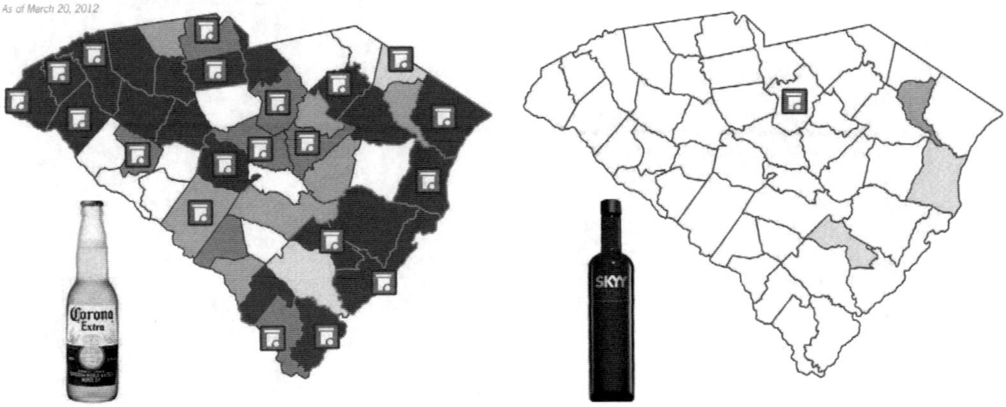

Earth911
The Recycling Authority

FACTOR: 5 CONTENT

HDPE JUGS vs. OIL BOTTLES

FLORIDA

DROPOFF AVAILABILITY
None
High

Curbside Available

As of March 16, 2012

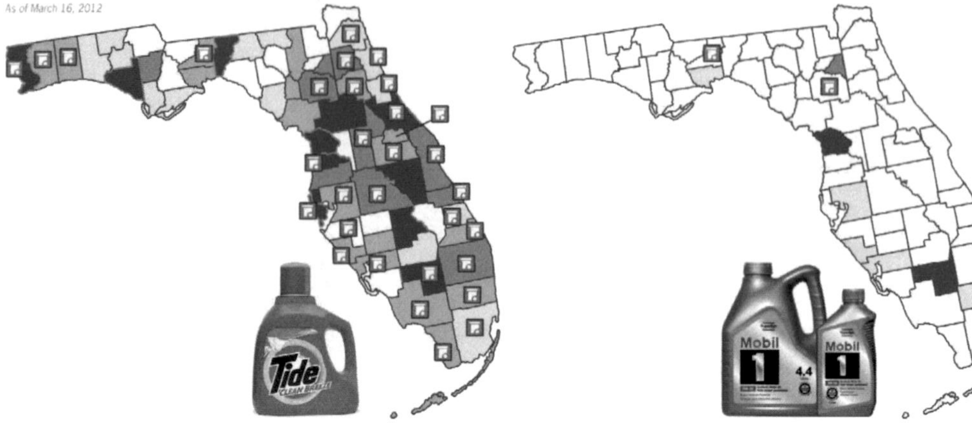

Earth911
The Recycling Authority

FACTOR 5: CONTENT

TIN CANS vs. AEROSOL CANS (full and empty)

WASHINGTON

DROPOFF AVAILABILITY
None
High

Curbside Available

As of March 16, 2012

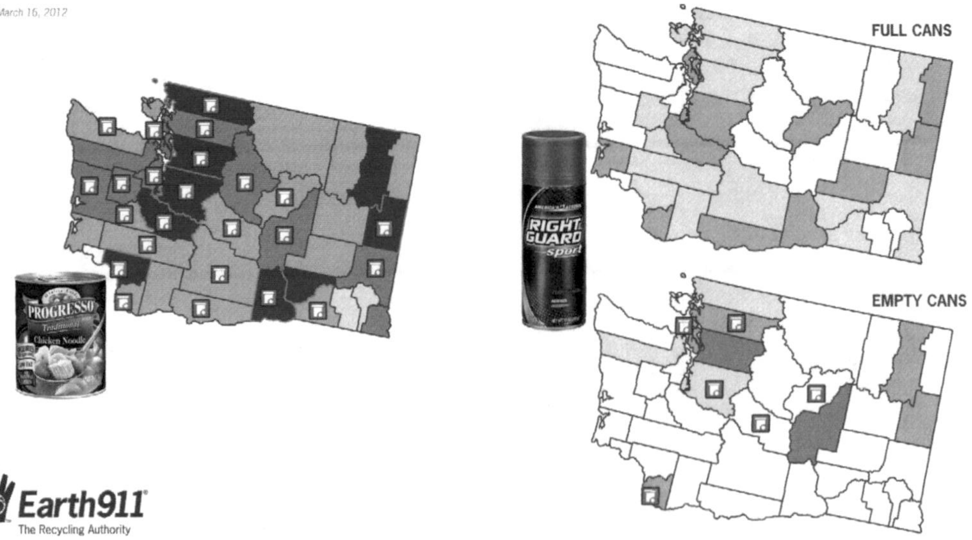

Earth911
The Recycling Authority

FACTOR 6: LEGISLATION

ALUMINUM BEVERAGE vs. ALUMINUM FOOD

NEW YORK

As of March 16, 2012

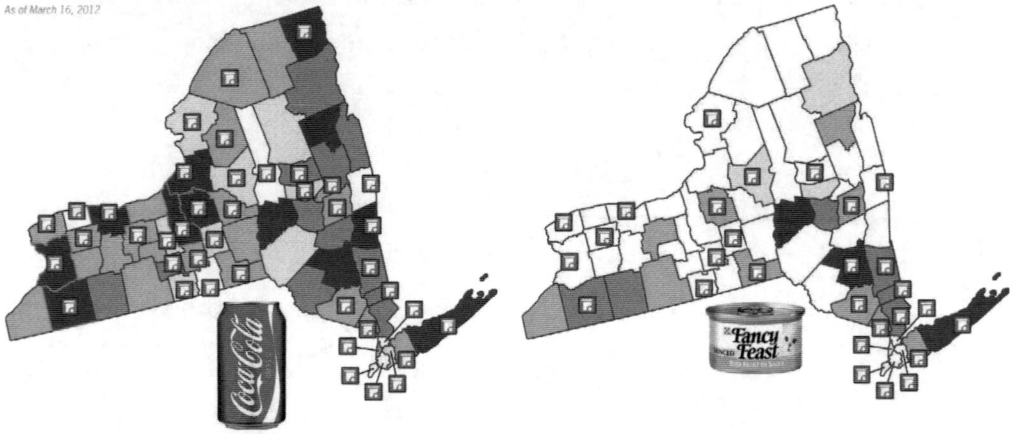

Earth911
The Recycling Authority

FACTOR 7: CATCH-ALL

ALUMINUM BEVERAGE CAN vs. CLEAR GLASS BEVERAGE BOTTLE

UTAH

As of March 16, 2012

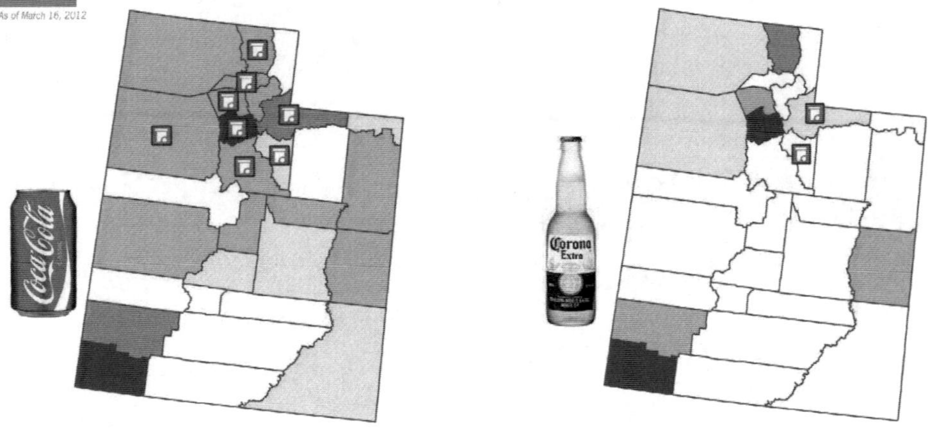

Earth911
The Recycling Authority

DEFINING "MATERIALS"

CONSUMER TERM		INDUSTRY TERM
▪ Cereal Box	=	▪ Paperboard
▪ Yogurt Tub		▪ #5 Rigid Plastic
▪ Milk Carton		▪ Gable-top Container
▪ Beer Bottle		▪ Brown Glass Container
▪ Soup Can		▪ Bi-metal Can
▪ Detergent Jug		▪ #2 Plastic Jugs - Colored

Earth911 The Recycling Authority

COMPLEXITIES OF MATERIALS

PACKAGING MATERIALS

Classification	Glass	Metal	Paper	Plastic
# of Materials	10	25	29	58
# of Keywords	165	425	314	1196
Average	16.5	17	10.8	20.6

NON-PACKAGING MATERIALS

Classification	Electronics	Garden	Hazardous
# of Materials	36	15	52
# of Keywords	413	137	802
Average	11.5	9.1	15.4

Earth911 The Recycling Authority

Earth911
The Recycling Authority

RECYCLING LOCATOR

1 Your Site 2 Earth911 Recycling Data 3 Your Recycling Locator

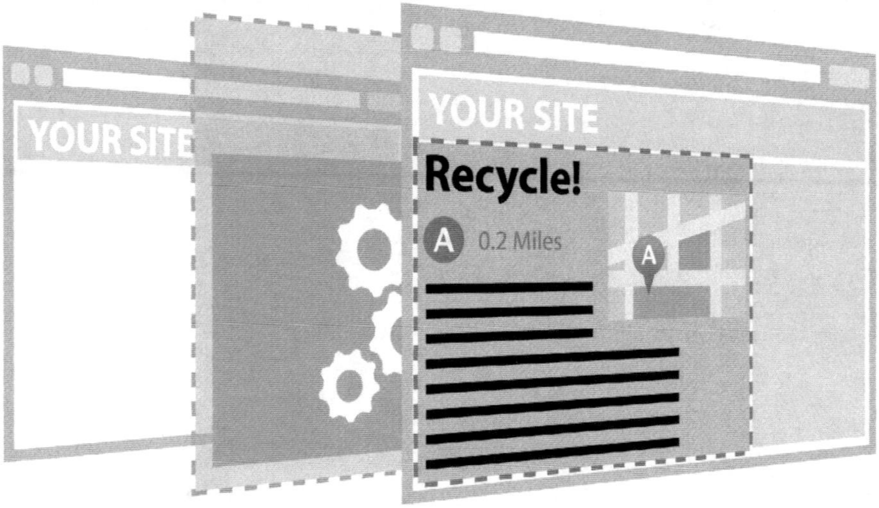

Earth911
The Recycling Authority

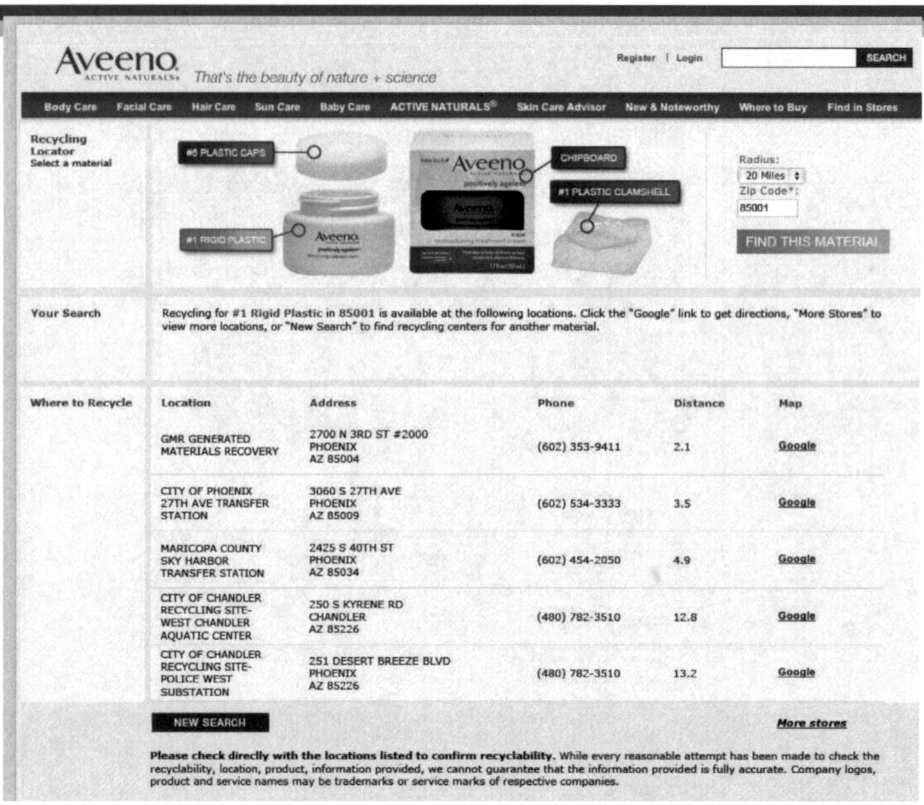

Factors to Consider in Package Design

- Resin or Grade
- Additives
- Form
- Color
- Content
- Legislation
- Other Factors

Earth911
The Recycling Authority

CONCLUSIONS

- A product is only recyclable if you can actually recycle it
- Small variables in materials can create HUGE differences in its recyclability
- Confusion is a leading reason for stream contamination
- Product specific instructions recommended to minimize consumer confusion and maximize your brand efforts

GOT SOMETHING YOU NEED TO RECYCLE? GIVE IT A TRY!
EARTH911.COM

Earth911
The Recycling Authority

THANK YOU

Sandra Keil
VP, Government Relations & Industry Affairs
Earth911, Inc.
skeil@earth911.com
(202) 577-1290

Earth911
The Recycling Authority

Plastics Resource Recovery

A piece of the sustainable packaging puzzle

Sustainable Packaging Symposium
Houston, April 4, 2012

Landfilling plastics = failure

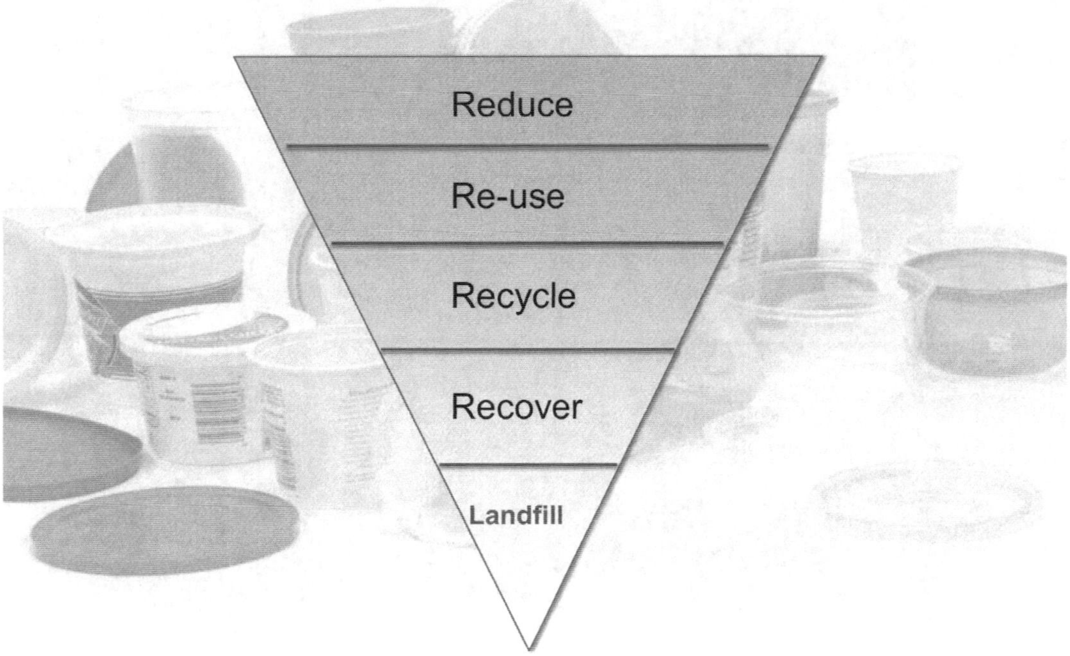

Plastics are borrowed energy

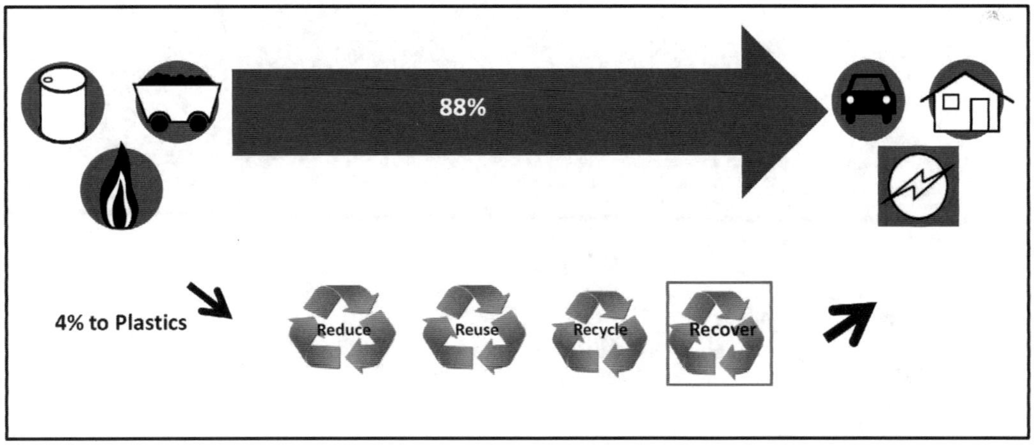

Sources: U.S. Energy Information Administration; American Chemistry Council

4

6

Life Cycle Inventory
Tuna in Can/Flexible Pouch

Table 1

TOTAL ENERGY, TOTAL SOLID WASTE, AND GREENHOUSE GASES
FOR 100,000 OUNCES OF TUNA CONSUMED

	Total Energy (MM Btu)	Total Solid Waste (lb)	Total Solid Waste (cu ft)	Greenhouse Gases (lb of CO2 equivalents)
Tuna Packaging Systems				
12-oz. Steel Can (1)	22.7	1,333	44.0	4,292
12-oz. Pouch	9.3	346	10.5	977
6-oz. Steel Can (1)	24.1	1,413	46.6	4,551
3-oz. Pouch	25.2	936	28.4	2,647
3-3-oz. Steel Cans in Paperboard Sleeve (1)	43.2	2,562	83.6	7,518
2-2.8-oz. Plastic Cups in Paperboard Sleeve	28.2	1,106	37.8	1,702

(1) End-of-life for the steel cans are modeled as 62% being recycled and 38% going to a landfill. The paper labels are assumed to be incinerated during steel recycling. Ash from the incineration of the labels is included in solid waste.
NOTE: The end-of-life for all other material is modeled as 80% going to a landfill and 20% combusted with energy recovery.

Source: Franklin Associates, a Division of ERG calculations using the Franklin Associates database and the U.S. LCI Database.

Eliminating plastic packaging would increase "footprint"

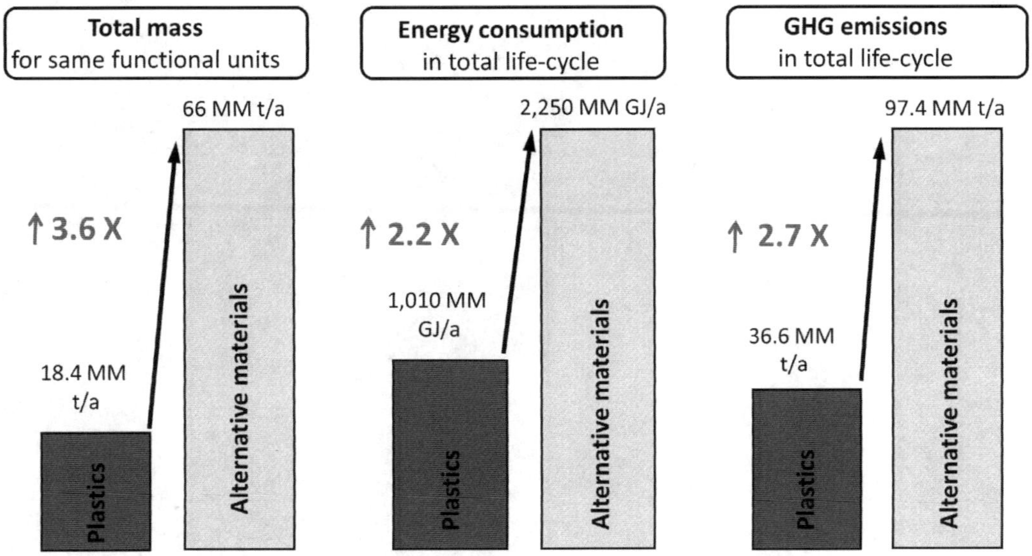

Source: Denkstatt GmbH Study - 'The Impact of Plastics on Life Cycle Energy Consumption and GHG Emissions in Europe", 2010. (EU 27 + 2)

More U.S. cities collecting rigid plastics

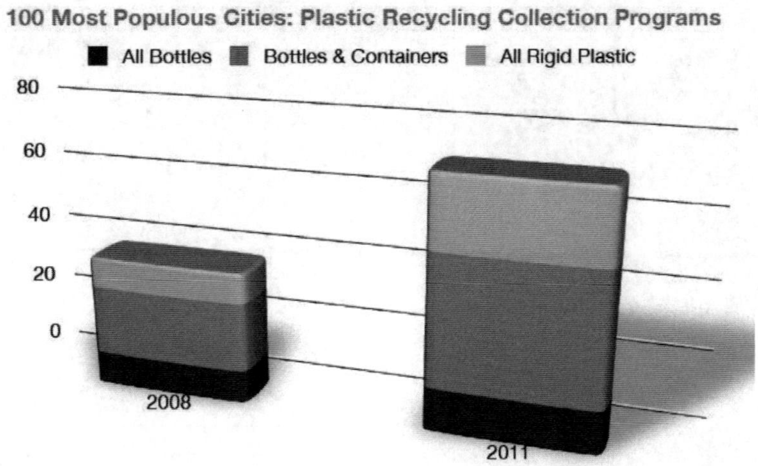

APR Rigid Plastics Recycling Program

Bale Definitions Subcommittee

Seven Types of Rigid Plastics Bales

	PET/HDPE Bottles	Bottles other than PET/HDPE	Non-bottle Containers	Bulky Rigid Plastics
All Rigid Plastics	X	X	X	X
Pre-picked Rigids		X	X	X
Bottles & Containers	X	X	X	
Household Containers		X	X	
Tubs & Lids			PE and PP only	
Olefin Bale	HDPE only	PE and PP only	PE and PP only	PE and PP only
Bulky Rigids				X

The Association of Postconsumer Plastic Recyclers
The voice of plastics recycling

Education
without
Numbers
Plastics
Recycling
Outreach

Plastic Bottles
caps on • no other plastic

All Plastic • No Bags
caps on • remove food
no polystyrene , bags, or wrappers

Plastic Containers
caps on • remove food
no polystyrene, bags, or wrappers

We are paying to bury energy

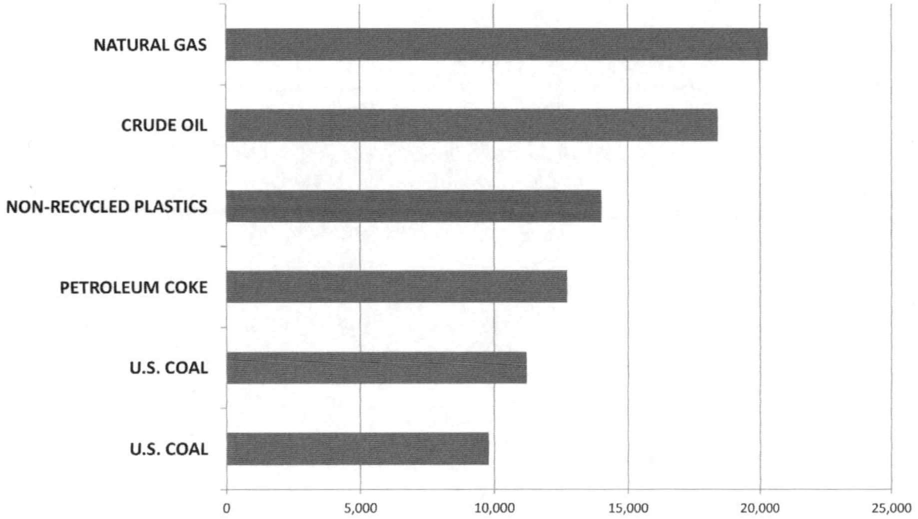

Natural Gas = 20,300 Btu/lb Crude Oil = 18,400 Btu/lb NRP = 14,000 Btu/lb Pet Coke = 12,700 Btu/lb Coal (High) = 11,200 Btu/lb Coal (Low) = 9,800 Btu/lb

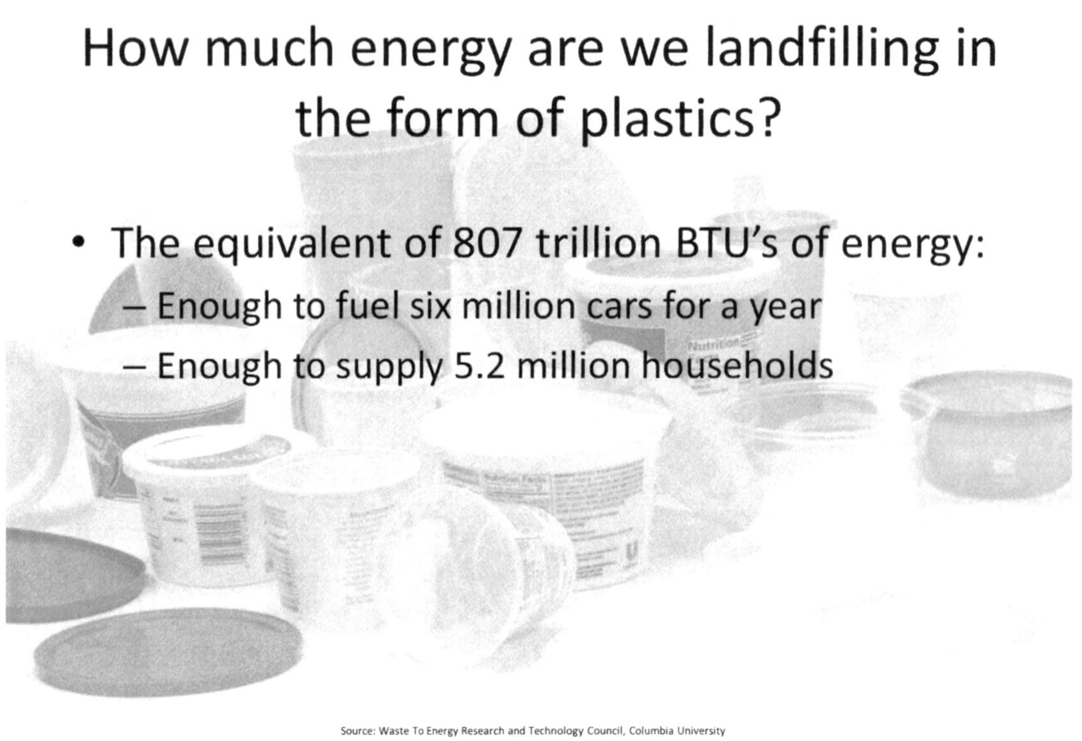
How much energy are we landfilling in the form of plastics?
• The equivalent of 807 trillion BTU's of energy:
– Enough to fuel six million cars for a year
– Enough to supply 5.2 million households
Source: Waste To Energy Research and Technology Council, Columbia University

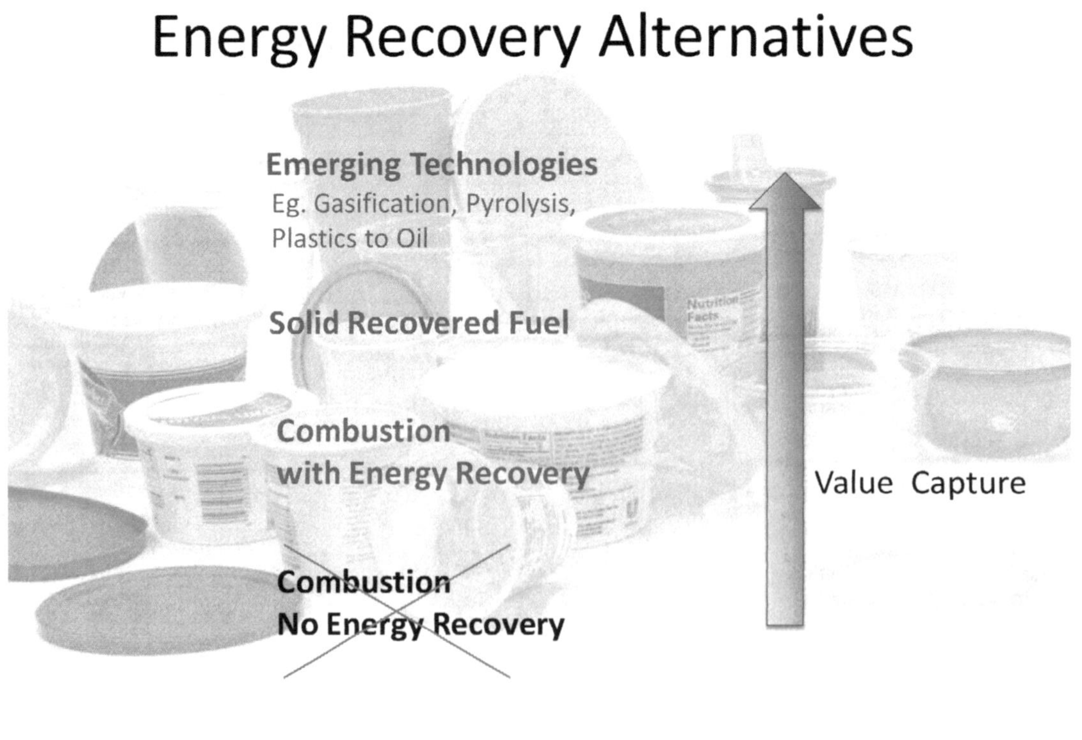
Energy Recovery Alternatives
Emerging Technologies
Eg. Gasification, Pyrolysis, Plastics to Oil
Solid Recovered Fuel
Combustion with Energy Recovery
Combustion No Energy Recovery
Value Capture

Waste to Energy

- Municipal Solid Waste (MSW) as fuel for electricity and steam generation
- 86 plants in the U.S.

Solid Recovered Fuel

- Plastics as solid fuel at cement kilns, industrial boilers, power plants.
- Commercially established
- New study from University of Texas

Gasification

Liquid Fuels from Municipal/Commercial Waste

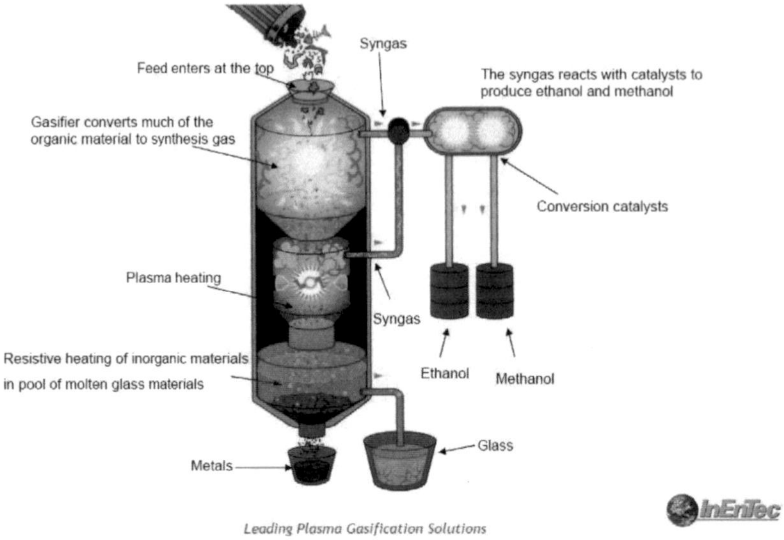

Pyrolysis – Plastics to Fuel

- Plastics as feedstock
- Pyrolysis process uses heat to de-polymerize plastics
- Products range from liquid fuels to industrial waxes

Opinion Leaders prefer Energy Recovery

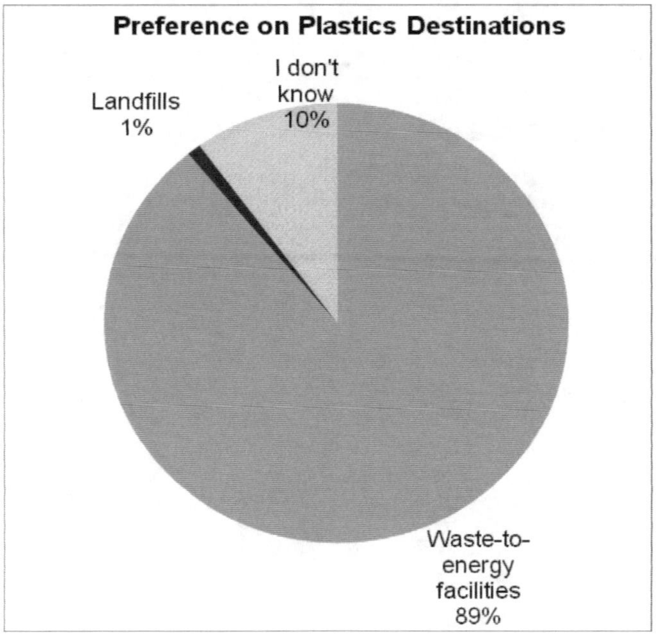

www.plasticsmakeitpossible.com

Information

Energy Recovery

Energy Recovery Council

Canadian Energy-From-Waste Coalition

Solid Waste Association of North America (SWANA)

Waste-to-Energy Research and Technology Council

recycle your plastics

- www.recycleyourplastics.org
- www.abagslife.com
- www.plasticsmarkets.org
- www.plasticbagrecycling.org
- www.marinedebrissolutions.org

THIRD OAK
ASSOCIATES INC.

Greg Wilkinson
gregw@thirdoakassociates.com
647.376.6008

Obstacles & Hurdles to WTE

Prof. Marco J. Castaldi
Department of Earth & Environmental Engineering
Henry Krumb School of Mines, Columbia University

Presentation to

SUSTAINABLE PACKAGING SYMPOSIUM 2012

12AIChE
Spring Meeting, Houston, TX

April 5th, 2012

WTERT
Waste-To-Energy Research and Technology Council

Wastes

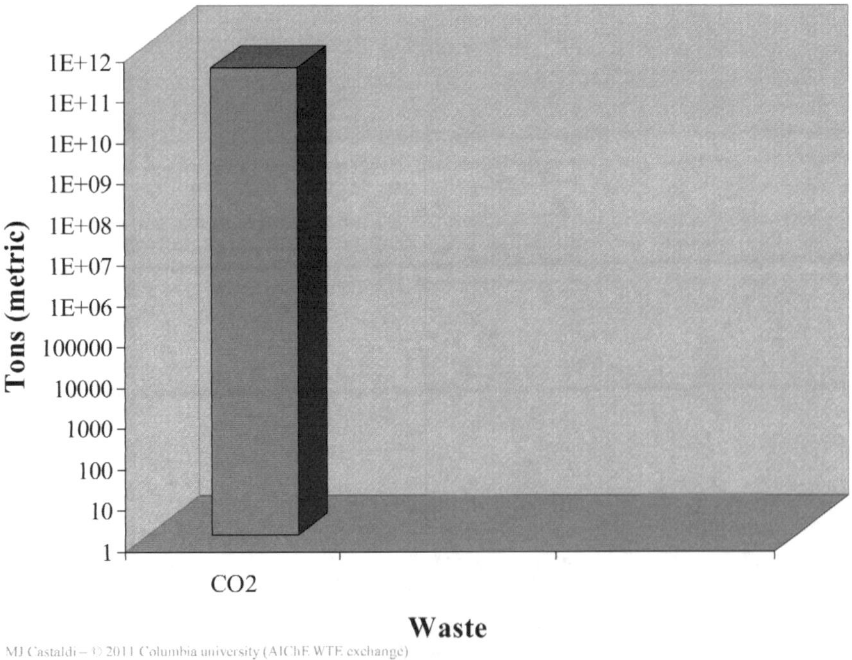

WTERT
Waste-To-Energy Research and Technology Council

The Future

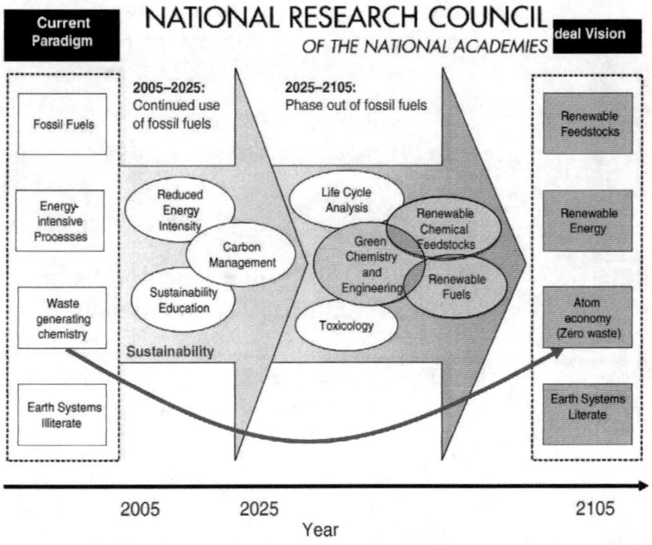

FIGURE ES-1 The Grand Challenges (ovals) for Sustainability (large arrows) that address the transition from current thinking to the ideal vision for the chemical industry over the next 100 years. See text for a more detailed description of figure and the Grand Challenges.

Source: *Committee on Grand Challenges for Sustainability in the Chemical Industry Board on Chemical Sciences and Technology Division on Earth and Life Studies*

- Thermal Conversions of waste are perfectly aligned with the ideal vision

WTERT Waste-To-Energy Research and Technology Council

MJ Castaldi © 2011 Columbia univ

Recommended Methods for Muncipal Solid Waste (MSW)

Expanded Hierarchy of Waste Management

(Themelis, GWMS, 2008)

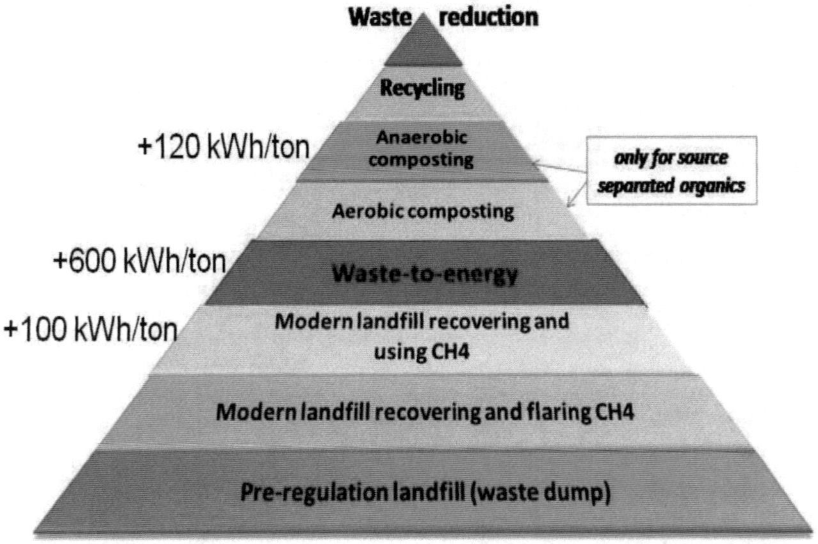

WTERT Waste-To-Energy Research and Technology Council

MJ Castaldi –

Two Primary choices for waste management

249 Million tons of trash (MSW) goes to landfills

Landfills

Renewable Energy Generated from Landfills - 5 billion kWh

Up to 100 kilowatt hours of electricity per ton of waste

Waste to Energy

29 Million tons of trash goes to EfW

Renewable energy generated from WTE Facilities - 15 billion kWh

Up to 700 kilowatt hours of electricity per ton of waste

These two options must co-exist for the foreseeable future

WIERT Waste-To-Energy Research and Technology Council

MJ Castaldi © 2011 Columbia university (AIChE WTE exchange)

WTE **Must** Increase

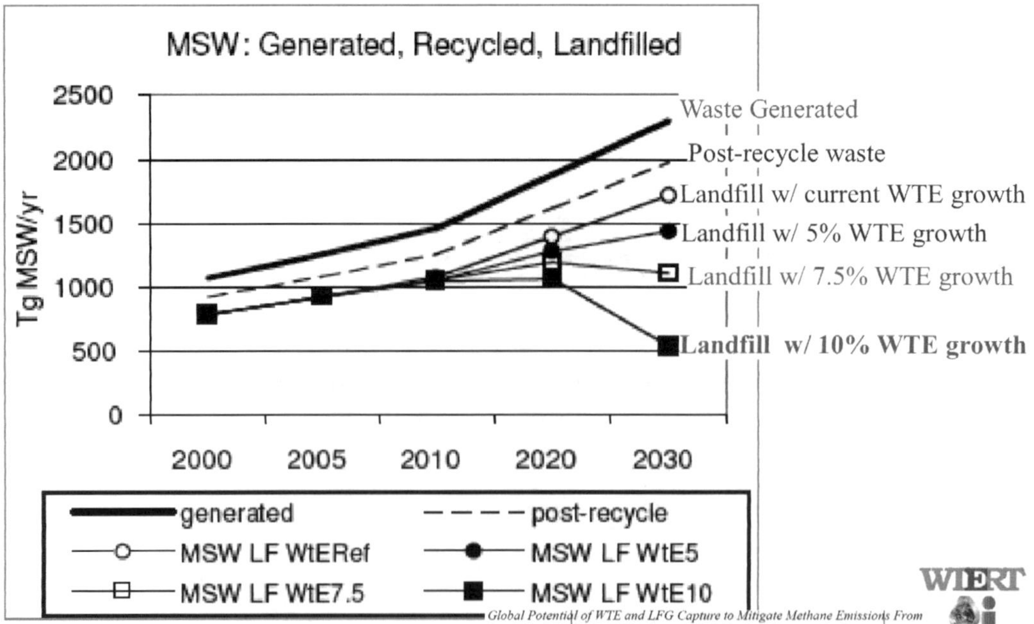

Global Potential of WTE and LFG Capture to Mitigate Methane Emissions From Landfills, 2000-2030, E. Matthews, NASA Goddard Institute for Space Studies, and N.J. Themelis (Sardinia Landfill Symposium, October 2007)

WIERT Waste-To-Energy Research and Technology Council

MJ Castaldi – © 2011 Columbia university (AIChE WTE exchange)

WTE & Landfill: A WM team

SUR Center for the study of Sustainable Use of Resources

Advancing the Goals of Sustainable Waste Management

Summary of the LF-WTE Meeting on Climate Impacts of U.S. Waste Management Industry, Washington, DC, Wednesday, January 28, 2009 (contains all comments and revisions submitted to SUR till March 27, 2009)

Objective: To identify areas of shared interest, agree upon data and information disseminated to the public regarding waste management, and establish better communication between the landfill and waste-to-energy industries, particularly with regard to the greenhouse gas footprint of the U.S. waste management industry.

Consensus reached

- Development of a consistent message by both landfill and WTE industry
- The WTE and landfill industry position should be clear in stating that the waste
- management industry must not be part of a "cap and trade" system,
- The % recovery of LFG at landfills is site-specific.
- *On an overall LCA basis, WTE is environmentally preferable to landfilling*
- The participants agreed to continue sharing scientific and policy information
- regarding GHG emissions and mitigation potential of WTE and landfill facilities.

WIERT Waste-To-Energy Research and Technology Council

Myths about WTE

1. Will Discourage Recycling
2. Will cost too much
3. Will cause health issues
    - Emissions are high (NOx, CO, Hg, PM, SO2)
    - Emitted Dioxins will kill you
4. environmental injustice
    - Only sited in poor areas of cities
    - Will cause more truck traffic
    - Will have odors
    - Ugly "factory-looking" structure, bring down the neighborhood
5. Better to focus on renewable fuels (CO2 reductions)
6. We need to be a "zero waste" society
7. Composting is better solution
8. Will encourage waste generation

WIERT Waste-To-Energy Research and Technology Council

Recycling

MJ Castaldi © 2011 Columbia university (AIChE WTE exchange)

WTERT
Waste-To-Energy Research and Technology Council

Economics

WTERT Waste-To-Energy Research and Technology Council

Cost of waste-to-energy

New plant or extension, EU Directive 2005
Construction cost, 700-900 USD/tonne/year

Treatment cost: ▪ Amortisation ▪ Operation ▪ Maintenance	=	USD/tonne	130
÷ Income from sale of electricity 2/3 MWh/t x 60 USD/MWh =	=	USD/tonne	÷ 40
÷ Income from sale of heat 2 MWh/t x 30 USD/MWh =	=	USD/tonne	÷ 60
Net treatment cost: Waste -to-energy	=	USD/tonne	30

Bettina Kamuk - NAWTEC 2006 May 2006 12

WTERT Waste-To-Energy Research and Technology Council

Energy Security and Economics

Security of EU energy supply: reducing dependence on fossil fuels

Waste to Energy plants in Europe

- Substitutes imported fossil fuels used by power plants.
- Supplies 7 million households with electricity
- 13.4 million households with heat.

The cost to avoid 1 tonne of CO_2

- Waste to Energy = 43€
- Biomass = 80€
- Photovoltaic >1000€

WTE is leading alternative energy technology when considering level of maturity, reliability and economics.

WTERT Waste-To-Energy Research and Technology Council

International Perspective

- **Europe**

  EU Landfill Directive

  65% reduction in landfilling of biodegradable Municipal Solid Waste

  Currently over 400 Waste-to-Energy Facilities

  2008 study estimated an additional 100 new facilities by 2012

- **China**

  Preferential feed-in electricity tariff

- **Kyoto Protocol**

  Recognizes Waste-to-Energy as an eligible offset in CDM protocol

- **Davos Report—The World Economic Forum**

  Identifies eight emerging clean energy sectors including wind, solar and Waste-to-Energy.

WTERT Waste-To-Energy Research and Technology Council

Health/Emissions

MJ Castaldi © 2011 Columbia university (AIChE WTE exchange)

WTERT Waste-To-Energy Research and Technology Council

Emissions from MSW incinerators 1930-1995

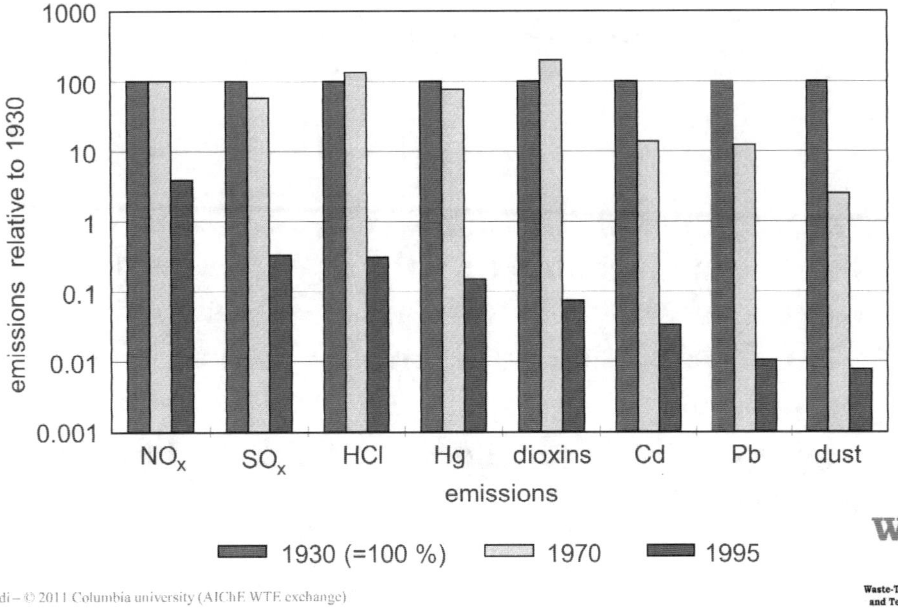

MJ Castaldi – © 2011 Columbia university (AIChE WTE exchange)

WTERT Waste-To-Energy Research and Technology Council

Example Waste-to-Energy Facility

Actual vs. Allowable Emissions

	Permit Limit	2003 – 2007 Actual Results	% Below Limit
Particulate (mg/dscm)	27	3.3	87.8%
Dioxin/Furan (ng/dscm)	30	1.7	94.3%
Mercury (μg/dscm)	28 – NY State (80 – Federal)	9.3	66.8% (88.4%)
Lead (μg/dscm)	440	7.6	98.3%
Cadmium (μg/dscm)	40	0.4	99.0%

Note: All concentrations corrected to 7% oxygen.

WTERT
Waste-To-Energy Research and Technology Council

Perception:

WTE facilities emit a lot of dioxins

Reality:

The total dioxin emissions from all U.S. WTEs have been estimated by EPA at 12 grams TEQ (toxic equivalent) of dioxins. However, before MACT regulation, they emitted about 10,000 grams TEQ

- **Major source of dioxins now: Backyard barrel burning (580 grams TEQ)**

WTERT
Waste-To-Energy Research and Technology Council

Dioxins Reality

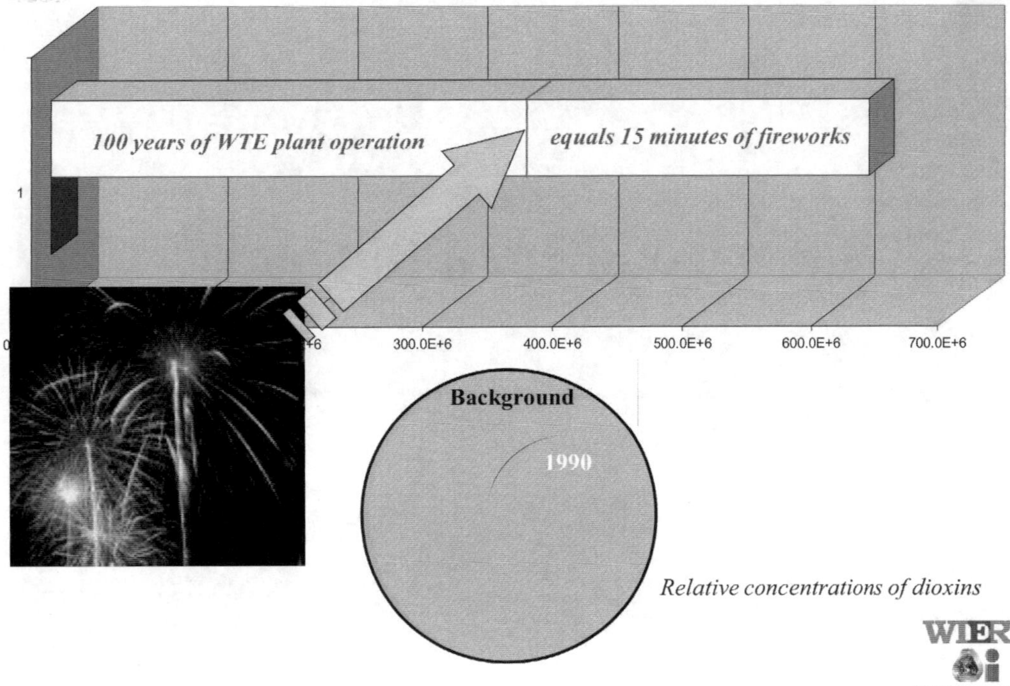

MJ Castaldi © 2011 Columbia university (AIChE WTE exchange)

WTERT Waste-To-Energy Research and Technology Council

Air Emissions - Dioxins

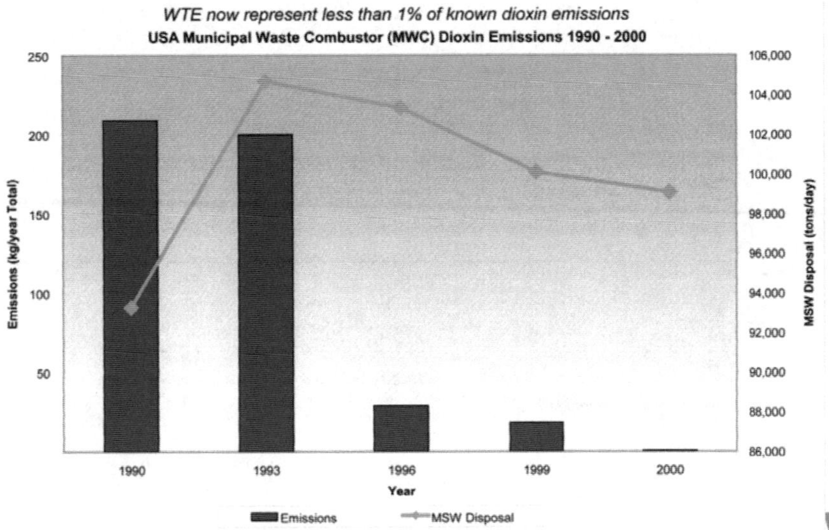

MJ Castaldi – © 2011 Columbia university (AIChE WTE exchange)

WTERT Waste-To-Energy Research and Technology Council

Environmental Injustice

MJ Castaldi © 2011 Columbia university (AIChE WTE exchange)

WIERT

Waste-To-Energy Research and Technology Council

Truck Traffic Avoided

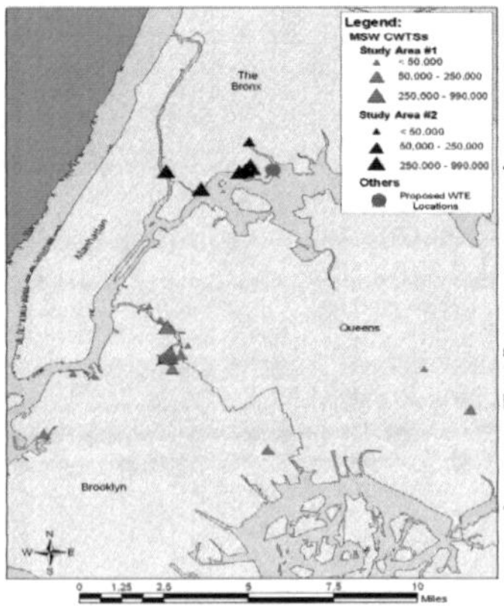

Table 2. Refuse Transport Distances Utilizing WTE A

Borough	One-way Travel from Source to Transfer Station to Landfill	One-way Travel from Source to Proposed WTE A	Decrease in One-way Travel (per truck) by Using WTE	Daily tonnage to WTE A from Borough (tpd)	Averted Truck Miles per day (Roundtrip)	Total Averted Truck Miles per year
taten Island	155.4	42.6	112.8	64.0	722	216,576
Manhattan	131.7	14.8	116.9	256.0	2,993	897,869
Bronx	134.4	9.0	125.4	1280.0	16,055	4,816,512
Brooklyn	142.4	27.8	114.7	1280.0	14,675	4,402,560
Queens	135.6	10.2	125.4	320.0	4,014	1,204,128
					Total:	11,537,645

Table 3. Refuse Transport Distances Utilizing WTE B

Address	Borough	One-way Travel from Source to Transfer Station to Landfill	One-way Travel from Source to Proposed WTE B	Decrease in One-way Travel (per truck) by Using WTE	Daily tonnage to WTE B from Borough (tpd)	Averted Truck Miles per day (Roundtrip)	Total Averted Truck Miles per year
Hopping Ave	Staten Island	155.4	29.7	125.7	64.0	805	241,363
South Street	Manhattan	131.7	6.0	125.7	256.0	3,218	965,453
West 261st & Palisades Ave	Bronx	143.4	17.9	125.4	1280.0	16,055	4,816,512
Bayview Avenue	Brooklyn	142.4	16.7	125.7	1280.0	16,091	4,827,264
Marathon Parkway	Queens	138.7	13.3	125.4	320.0	4,014	1,204,128

MJ Castaldi – © 2011 Columbia university (AIChE WTE exchange)

Waste-To-Energy Research and Technology Council

Development of MSW incineration 1900 - 2000

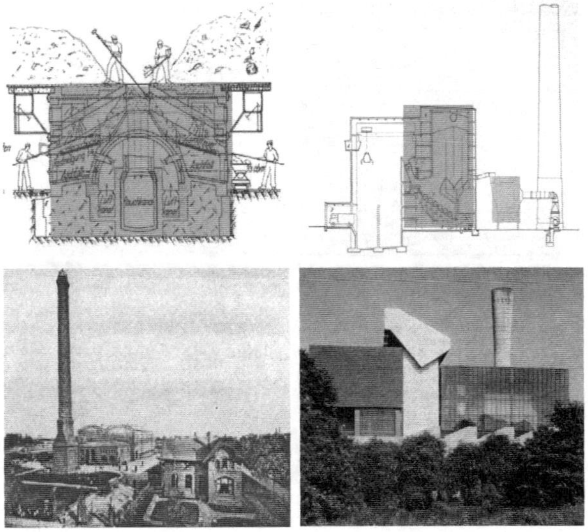

1896 1970

WTE Facilities Today

MJ Castaldi © 2011 Columbia university (AIChE WTE exchange)

Waste-To-Energy Research and Technology Council

Renewables

MJ Castaldi – © 2011 Columbia university (AIChE WTE exchange)

WTERT

Waste-To-Energy Research and Technology Council

A Renewable Resource

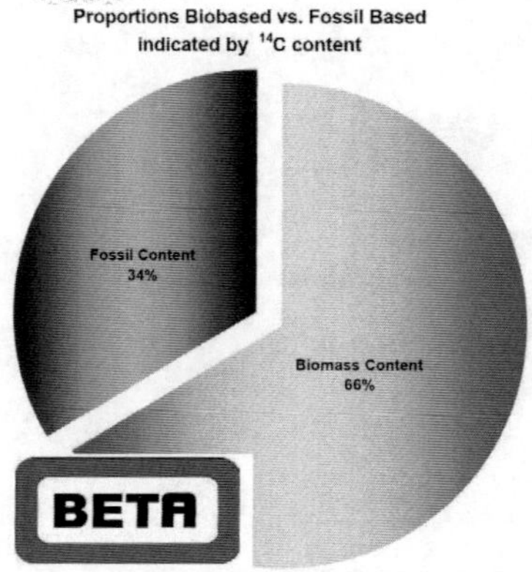

Table 3. Characterization of U.S. MSW by USEPA [7]

Biomass components	%	Petrochemical components	%
Paper/board	36.2	Plastics	11.3
Wood	5.8	Rubber, nylon, other textiles.*	3.7
Yard trimmings	12.1		
Food scraps	11.7		
Textiles (cotton, wool, leather)*	3.7		
Total biomass	69.5%	Total fossil-based	15.0%

**Rubber, leather and textiles category of USEPA 7.4%) were assumed to be divided equally between natural and man-made products*

- 3 random grab samples
- Results show biomass content
  66%, 68% and 66%
- Biocarbon emission = 0.19 MTCE/ton MSW
- Fossil based emission = 0.08 MTCE/ton MSW
- 10% - 36% reduction of CO_2 emissions compared to landfill

WTERT Waste-To-Energy Research and Technology Council

Composting

WTERT Waste-To-Energy Research and Technology Council

Composting does not amount to much

P. Ulloa, "Overview of Food Waste Composting in the U.S." Internal Report, Earth Engineering Center, Columbia University, July 2008.

Table 1. Data and Estimates of Residential Food Waste in the Unites States

MSW Generated[1]	tons	387,855,461
Total Quantity of Organics Composted (including yard trimming, MSW, and SSO)[1]	tons	20,368,139
Percentage of Organics Composted of the total MSW Generated	%	5.3%
Percentage of Food Waste[2]	%	12.4%
Estimated Quantity of Food Waste	tons	48,094,077
Reported Quantity of Food Waste Composted[2]	tons	680,000
Proportion of Food Waste Captured	%	1.4%

[1] Simmons et.al., BioCycle, SOG 2006; [2] EPA, 2006

WIERT
Waste-To-Energy Research and Technology Council

Gasification

WIERT
Waste-To-Energy Research and Technology Council

Pyrolysis, Gasification or Combustion

- Normally no air
- Only heat (external or internal)
- Want liquid, Gases not desired
- Pollutants in reduced form (H_2S, COS)
- High Char
- **Scale: ~ 10 tons/day**

- Sub stoichiometric air
- Lower total volumetric flow
- Lower fly ash carry over
- Pollutants in reduced form (H_2S, COS)
- Char @ Low T
- Vitrified Slag @ high T
- **Scale: ~ 100 tons/day**

- Excess air
- Higher volumetric flowrate
- Fly ash carry over
- Pollutants in oxidized form (SO_x, NO_x, etc)
- Bottom ash
- **Scale: ~ 1500 tons/day**

No additional Oxygen (only heat) Unconverted solid will remain!

Some additional Oxygen (or air) Heat added or comes from reactions

Much additional Oxygen (or air) Heat comes from reactions

WTERT Waste-To-Energy Research and Technology Council

- ***Most HC fuels ($C_xH_yO_zN_aS_b$) exist above lines***
- ***Need oxygen source or hydrogen to get below C(s) line (CO_2, O_2, air H_2O, H_2)***

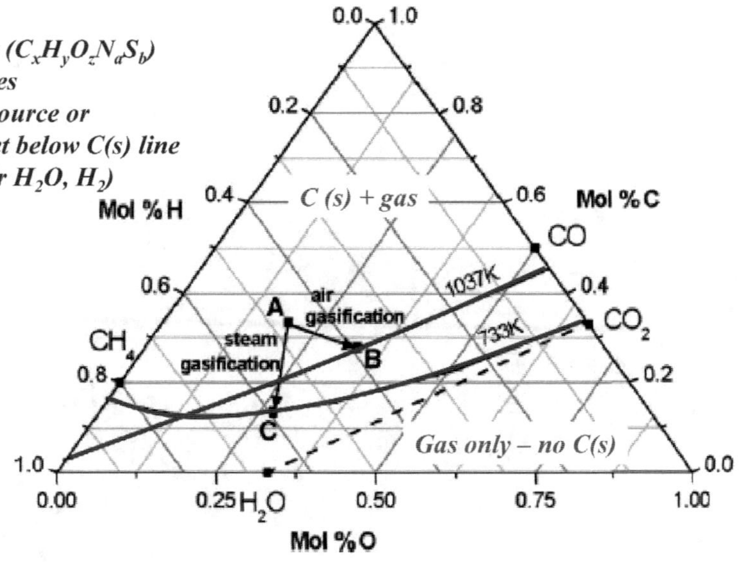

Fig. 1. Molar triangular diagram indicating (A) biomass feed, (B) biomass in equilibrium with air at carbon boundary, (C) biomass in equilibrium with steam of 500 K at carbon boundary.

WTERT Waste-To-Energy Research and Technology Council

M. J. Prins et al. / Chemical Engineering Science 58 (2003) 1003–1011

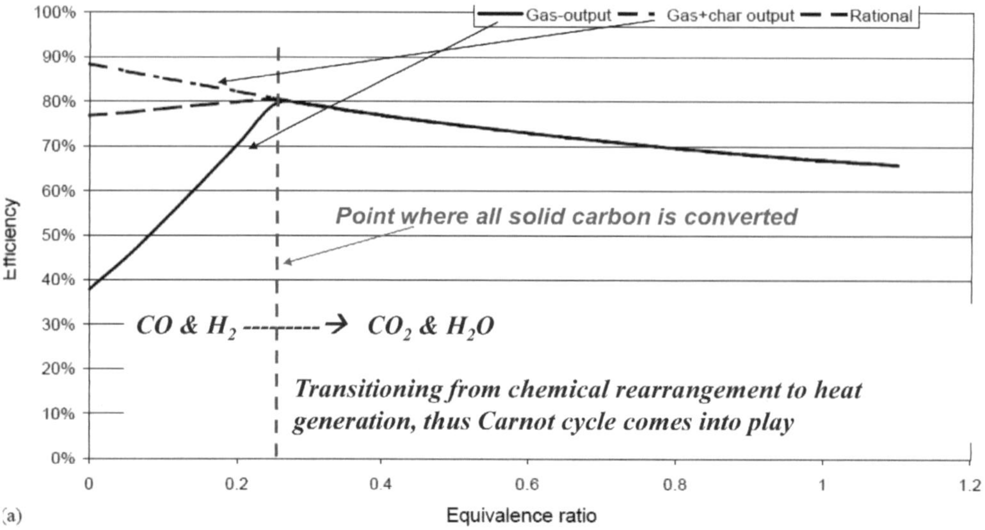

Typical Waste-to-Energy Plant

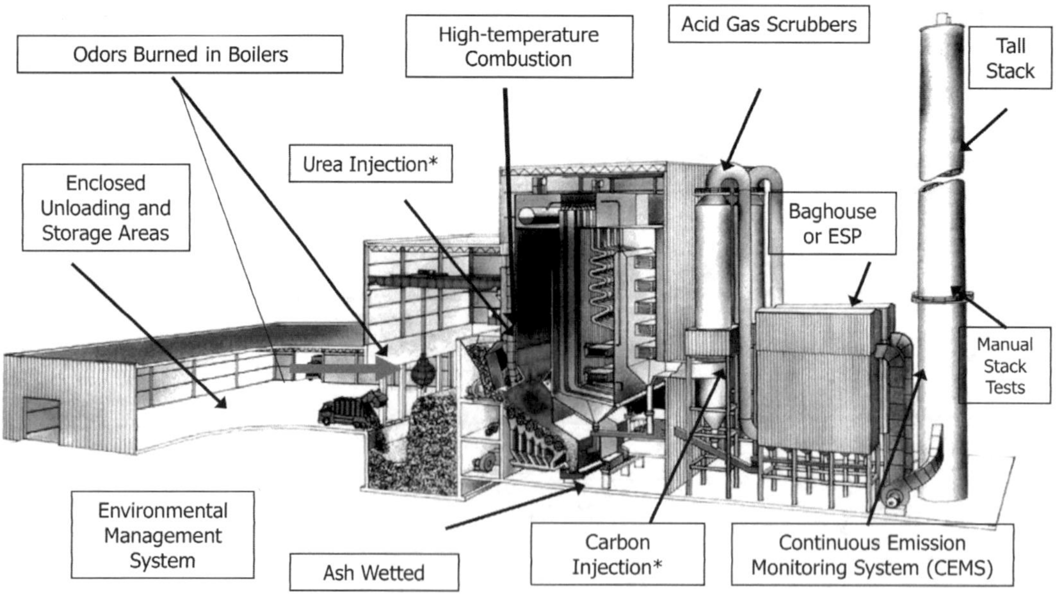

* Some Plants

The Thermoselect Process (Interstate Waste Technologies, U.S.)

THERMOSELECT

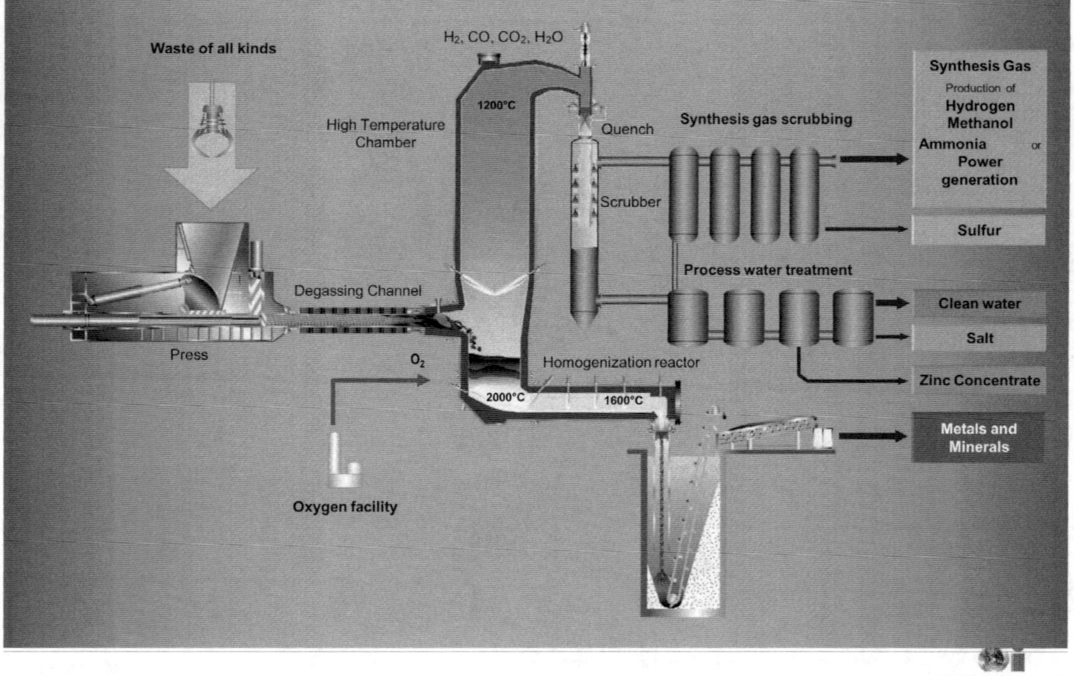

Waste-To-Energy Research and Technology Council

Ebara Fluidized-bed TwinRec/TIFG Technology

Process Flowsheet

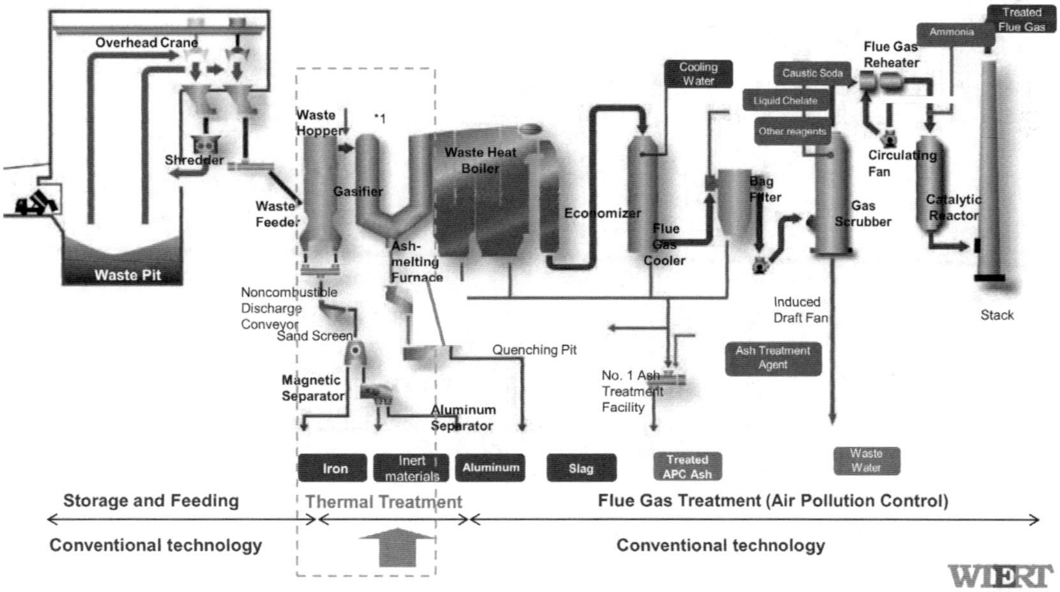

WTERT

Waste-To-Energy Research and Technology Council

The Nippon Steel WTE Process

Basic Flow Chart of Direct Melting System

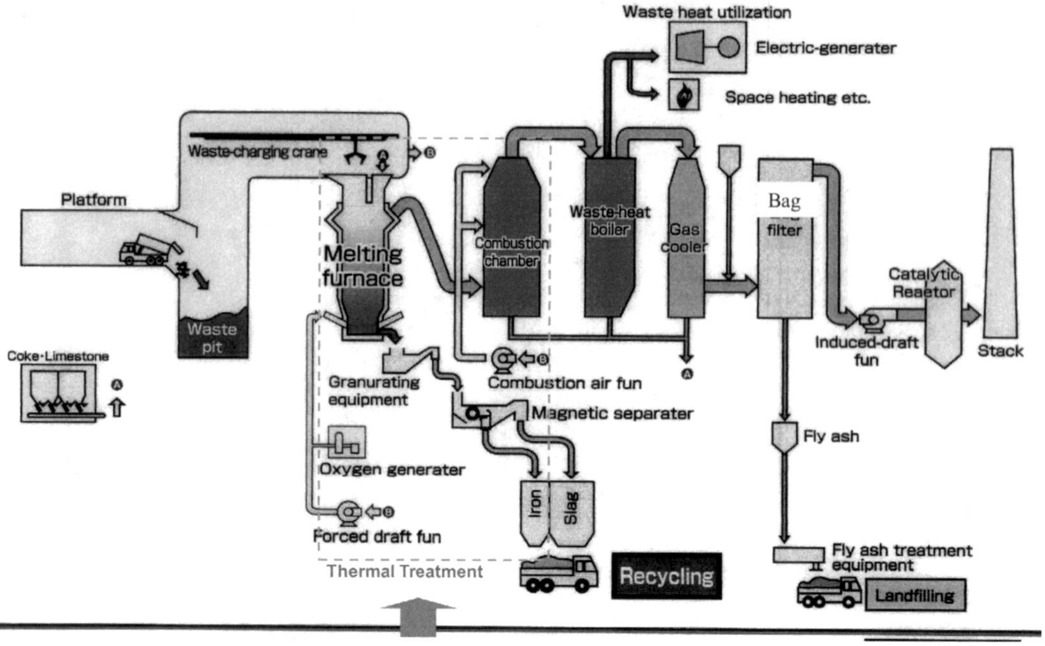

MJ Castaldi © 2011 Columbia university (AIChE WTE exchange)

Waste-To-Energy Research and Technology Council

Chemicals From Waste

- Military MISER program
  - Trash/Biomass/Solid hydrocarbons to fuels
- American Chemical Society (ACS)
  - Letters to the editor – "chemicals from waste"
    - C&EN April 2006
- Discover Magazine –
  - "DATA" Section : The Ultimate Garbage Disposal
    - How to turn trash into clean energy – Geoplasma Unit
      - 160 MW by 2009, St. Lucie County, Florida
- EnerChem/City of Edmonton - 2008

WTERT

Waste-To-Energy Research and Technology Council

MJ Castaldi – © 2011 Columbia university (AIChE WTE exchange)

Plasma Gasification Field System Summary

Technology	Energy (kWh/ton)	Capital Costs ($/ton)
InEnTec	530	~77 (est)
Alter NRG	617	81
Europlasma	605	86
Plasco	530	86
Grate WTE (US avg)	550	60

Public Perception Changing for the positive

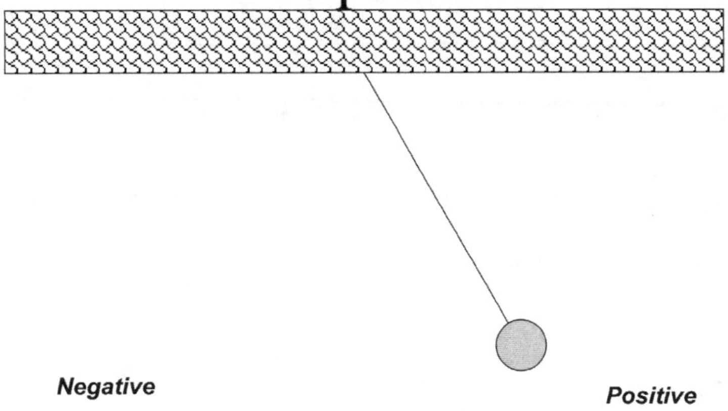

WasteManagement & InEnTec

S4 Energy Solutions

- Joint Venture Formed to Commercially Deploy Gasification of Waste
- 50/50 Joint Venture Between WM and InEnTec
- Venture Officially Launched in February 2009
- JV to Operate in Collaboration with, but Independent of JV Partners

S4 Energy SOLUTIONS

WIERT Waste-To-Energy Research and Technology Council

Environment

Europe Finds Clean Energy in Trash, but U.S. Lags

By ELISABETH ROSENTHAL
Published: April 12, 2010

HORSHOLM, Denmark — The lawyers and engineers who dwell in an elegant enclave here are at peace with the hulking neighbor just over the back fence: a vast energy plant that burns thousands of tons of household garbage and industrial waste, round the clock.

SIGN IN TO RECOMMEND
TWITTER
SIGN IN TO E-MAIL
PRINT
SINGLE PAGE
REPRINTS
SHARE

Enlarge This Image

Johan Spanner for The New York Times

A plant in Horsholm, Denmark, uses new technology to convert trash into energy more cleanly.

Far cleaner than conventional incinerators, this new type of plant converts local trash into heat and electricity. Dozens of filters catch pollutants, from mercury to dioxin, that would have emerged from its smokestack only a decade ago.

In that time, such plants have become both the mainstay of garbage disposal and a crucial fuel source across Denmark, from wealthy exurbs like Horsholm to Copenhagen's downtown area. Their use has not only reduced the country's energy costs and reliance on oil and gas, but also benefited the environment, diminishing the use of landfills and cutting carbon dioxide emissions. The plants run so cleanly that many times more dioxin is now released from home fireplaces and backyard barbecues than from incineration.

Multimedia

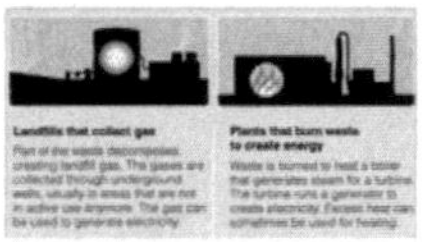

Graphic

Two Approaches to Waste

MOST POPULAR

E-MAILED | BLOGGED | SEARCHED | VIEWED

1. Nicholas D. Kristof: New Alarm Bells About Chemicals and Cancer
2. Signs of Neanderthals Mating With Humans
3. Paul Krugman: A Money Too Far
4. Movie Review | 'Babies': Awwwwww, Ewwww, Who's a Good Baby?
5. A Sampling of Chinglish
6. Private School Screening Test Loses Some Clout
7. Facebook Glitch Brings New Privacy Worries
8. David Brooks: Leading With Two Minds
9. Teenage Insults, Scrawled on Web, Not on Walls
10. Op-Ed Contributor: The Last Days of the Dragon Lady

Go to Complete List »

HOME PAGE | TODAY'S PAPER | VIDEO | MOST POPULAR | TIMES TOPICS

Log In | Register Now | TimesPeople

The New York Times
Friday, May 7, 2010

Opinion

Search All NYTimes.com

WORLD | U.S. | N.Y. / REGION | BUSINESS | TECHNOLOGY | SCIENCE | HEALTH | SPORTS | OPINION | ARTS | STYLE | TRAVEL | JOBS | REAL ESTATE | AUTOS

EDITORIALS COLUMNISTS CONTRIBUTORS LETTERS THE PUBLIC EDITOR GLOBAL OPINION

ROOM FOR DEBATE — A Running Commentary on the News

April 13, 2010, 6:16 PM

Should the U.S. Burn or Bury Its Trash?

By THE EDITORS

Robert F. Bukaty/Associated Press

A landfill in Bath, Maine.

Updated, April 14, 3:30 p.m. | Neil Seldman of the Institute for Local Self-Reliance joins the discussion.

Search This Blog

Previous post: Why Do Educated People Use Bad Words?

Next post: What Tea Party Backers Want

Recent Discussions

- Rules, Revolving Doors and the Oil Industry (58)
- Who Benefits When Airlines Merge? (49)
- Times Square, Bombs and Big Crowds (85)
- Should Laws Push for Organ Donation? (203)
- When Is It Smart to Abandon Your Party? (55)
- What the Spill Means for Offshore Drilling (326)
- Who Should Be the Next Justice? (99)

Subscribe

Room for Debate RSS

FOLLOW US ON Twitter

Archive

Select Month

The Gulf of Mexico Oil Spill

Revolving Doors and the Oil Industry

What the Spill Means for Oil Drilling

- 104 (57%) comments supported WtE.
- 43 (24%) comments opposed WtE.
- 34 (19%) comments unclear.
- 181 comments in total.

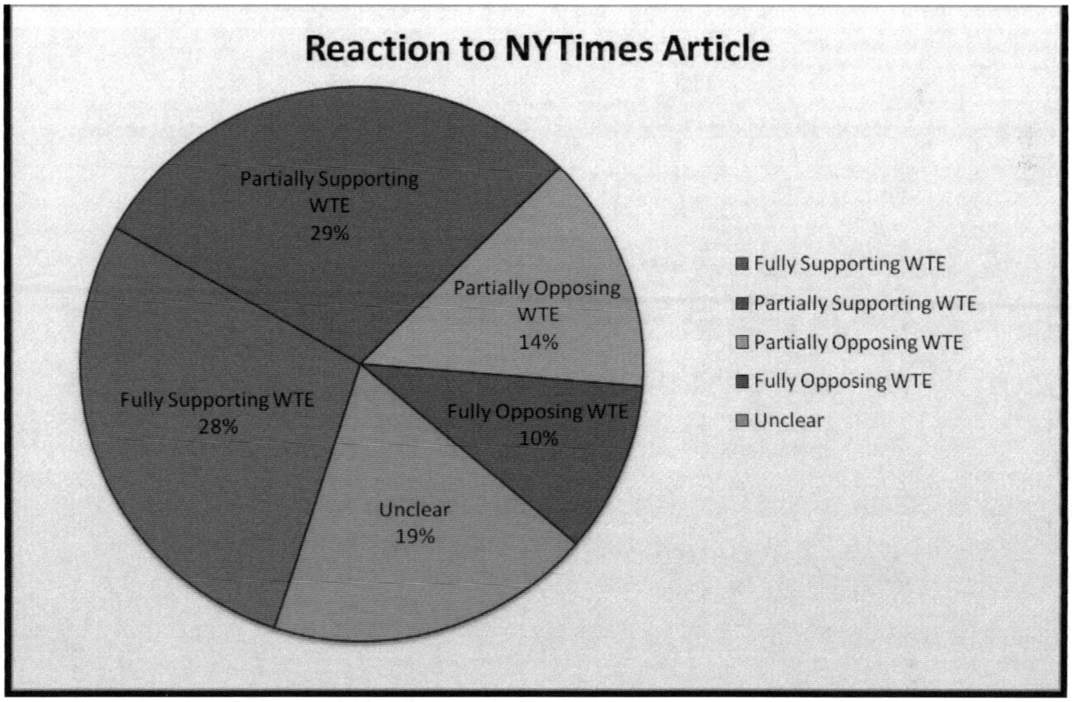

New Construction!!

Friday June 4, 2010

SITE SEARCH GO

News Archive Search

WASTE & RECYCLING NEWS

www.wasterecyclingnews.com

...and powerfully.

Open for Business! Waste & Recycling News' Brand New Online Buyers Guide

EDITORIAL

Home/News

Data & Research Center NEW

Landfill Report

Residential Recycling Report

Scrap Report

Photo Gallery

Special Reports

Departments

Opinion

Archives

Wheelabrator to build proposed U.S. WTE facility

Feb. 4 -- Wheelabrator Technologies Inc. has been tapped to build a proposed waste-to-energy facility in Maryland, the first new U.S. plant in more than a decade.

The Northeast Maryland Waste Disposal Authority selected the company as the preferred vendor to build the plant pending approval from Frederick and Carroll counties, which the plant would serve. The proposed plant would be located in Frederick County. It would be the first new greenfield waste-to-energy facility built in the United States in more than a decade.

The 55-megawatt plant would have the capacity to dispose of 1,500 tons of waste per day and generate enough electricity for some 60,000 homes. It will employ about 50 full-time employees. Wheelabrator expects to complete the facility in 2014, if approved.

The NMWDA will own the facility, which will serve Frederick and Carroll counties under a long-term service agreement.

Contact Waste & Recycling News reporter Joe Truini at 330-865-6166 or jtruini@crain.com

Covanta Energy plans to build an approximately $575 million plant to convert waste to energy Wales (UK).

The plant will supply electricity for up to 180,000 homes in Wales, create 600 jobs, and reduce the need for municipal, commercial and industrial waste landfill sites. The plant will be on an existing railway line that will link it to rail-operated waste transfer stations across Wales. Transporting waste by rail will greatly reduce the need for road haulage of waste. Almost 500 jobs will be created during construction of the plant, which will have a full time staff of approximately 100 when it is operational. Covanta Energy expects to have the plant operational by 2014, subject to approval by local authorities. The company is also investigating the possibility that residents and businesses in neighboring communities could buy electricity at significantly below market rates, and that lower cost energy and waste heat can be made available to attract other investors to the area.

WTERT Waste-To-Energy Research and Technology Council

WTE will increase

- 2011 world MSW generating ~ 2 billion tons
- ~ 800 thermal WTE plants currently
  - Operating in nearly 40 countries
  - This is 11% vs 70% landfilled
- WTE expected to increase
  - to 283 terawatt hours by 2022 from 221 in 2010
  - Global market for WTE technologies ~$6.2 billion in 2012 and grow to $29.2 billion by 2022.

WTERT Waste-To-Energy Research and Technology Council

Boundaries and interfaces of WTE cut across all sustainable fronts

- ***WTE conserves fossil fuels by generating electricity.*** *(Energy)*
  - 1 ton of waste combusted = 45 gallons of oil or 0.28 tons of coal
  - Most WTE facilities in U.S. process between 500 and 3,000 tons of waste per day
  - Electricity for 2.8 million homes
- **WTE facilities process 14% of the MSW in the United States.** **(Health)**
  - Trash-disposal needs about than 37 million people
- **WTE facilities meet some of the world's most stringent standards.** **(Environmental, local)**
  - Achieved compliance with new Clean Air Act pollution control standards in 2000
  - EPA data :dioxin emissions now account for less than 0.5% of dioxin emissions
- ***WTE facilities reduce greenhouse gas emissions.*** *(Climate, global)*
  - EPA estimates :WTE facilities prevent 33 million metric tons of CO_2 per year avoided
- ***WTE facilities save real estate.*** *(Land)*
  - They reduce the space required for landfills by about 90%
- ***WTE is compatible with recycling.*** *(Resource Minimization)*
  - Communities served by WTE recycle 35% of their trash, compared to 30% for the general population.
  - Annually removes more than 700,000 tons of ferrous materials
  - ~ 3 million tons of WTE ash reused as landfill cover, roadbed, or building material.
- ***WTE facilities provide economic benefits.*** **(Economic)**
  - WTE is a $10 billion industry employs ~ 6,000 American workers annual wages ~ $400 million

MJ Castaldi © 2011 Columbia university (AIChE WTE exchange)

Waste-To-Energy Research and Technology Council

WTERT Waste-to-Energy Research and Technology Council

Advancing the Goals of Sustainable Waste Management

What is WTERT

The Waste-to-Energy Research and Technology Council (WTERT) is a top-tier-technical group that brings together engineers, scientists, and managers from industry, universities, and government with the objective of advancing the goals of sustainable waste management globally.

http://www.seas.columbia.edu/earth/wtert/index.html

- **Industry Award:** *World's Best Waste-to-Energy facilities*:
  - **2006 – ASM Brescia, Italy**
  - **2004 – Martin GmbH, Germany**
- **Education Award:** *Person who has advanced Integrated Waste Management globally:*
  - **2006 – Prof. Paul H. Brunner,** Technical University of Vienna
  - **2004 – Prof. George Tchobanoglous,** University of California- Davis

- State University of New York, Marine Sciences Research Center (U.S.)
- Delft University of Technology, Dept. of Applied Earth Sciences (The Netherlands)
- Sheffield University Waste Incineration Centre (United Kingdom)
- National Tech. University of Athens (Greece)
- University of Patras (Greece)
- Institute for Thermal Power Engineering of Zhejiang University (China)
- Collage of Mechanical and Energy Engineering Zhejiang University (China)
- Chongqing Waste to Energy Technology Research Institute (China)

MJ Castaldi © 2011 Columbia university (AIChE WTE exchange)

Thank you for your attention

...for more information

http://www.seas.columbia.edu/earth/wtert/

http://www.cclabs.org

WTERT

Waste-To-Energy Research and Technology Council

MJ Castaldi © 2011 Columbia university (AIChE WTE exchange)

V. Environmental: SO2 LCA Comparison[2]

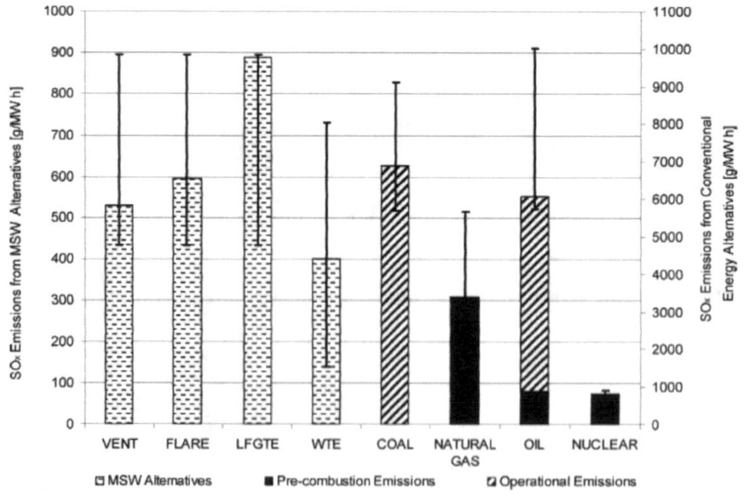

FIGURE 3. Comparison of sulfur oxide emissions for LFGTE, WTE, and conventional electricity-generating technologies (Tables S5–S8, Supporting Information, include the full data set).

MJ Castaldi – © 2011 Columbia university (AIChE WTE exchange)

Waste-To-Energy Research and Technology Council

V. Environmental: NOx LCA Comparison[2]

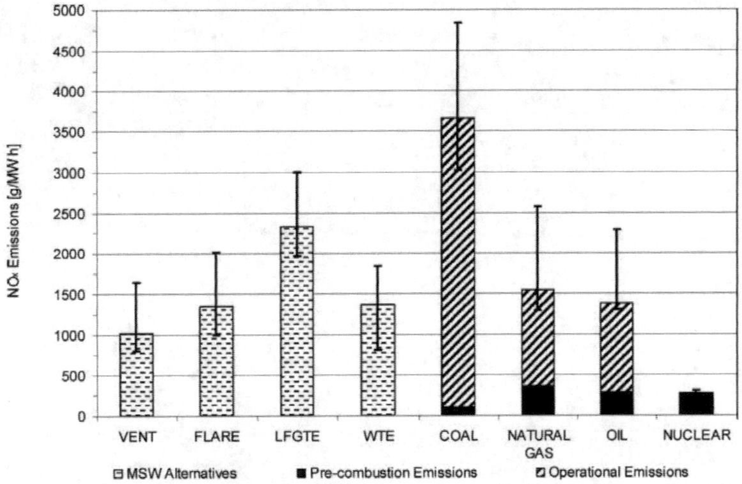

FIGURE 4. Comparison of nitrogen oxide emissions for LFGTE, WTE, and conventional electricity-generating technologies (Tables S5–S8, Supporting Information, include the full data set).

Waste-To-Energy Research and Technology Council

Waste-To-Energy Research and Technology Council

COALITION *for* RESOURCE RECOVERY

GLOBAL GREEN USA

Post-Use Optimization: Resource Recovery as a Key Component of Packaging Sustainability

Annie White

April 5, 2012

About Global Green USA

- Global Green: American chapter of Green Cross International
- Founded by President Mikhail S. Gorbachev to foster a value shift to a sustainable and secure future by reconnecting humanity with the environment
- Global Green creates sustainable urban environments that combat global warming through a unique cross-cutting approach that merges innovative research, technical assistance, cutting-edge community based projects and targeted education and outreach
- 501c3 non-profit organization

COALITION *for* RESOURCE RECOVERY

GLOBAL GREEN USA

Who We Are:

Global Green's Coalition for Resource Recovery (CoRR)

- Industry working group dedicated to combating climate change and transforming waste to assets
- Conduct pilots and related research to identify and accelerate development of scalable, transferable waste diversion programs and technologies.
- Conduct targeted education & outreach to NGOs, trade associations & municipalities to facilitate transferability of identified solutions

COALITION *for* RESOURCE RECOVERY

A Coalition of Members

Action Env'l Group	First Fiber	Pret A Manger
BASF	Green Bay Packaging	Spectra-Kote
Bemis	HAVI Global Solutions	Starbucks
Chemol	IESI/Progressive	Transtech, Inc.
Clean River Systems	Imerys	Ulterion
DBB Partners	Interstate Container	Waste Management
Design & Source	LBP Manufacturing	Wastequip
Duro Bag	Pratt Industries	Western Michigan Univ.

COALITION *for* RESOURCE RECOVERY

Agenda

- New York City recent developments
- Wholesale packaging
- Foodservice packaging

COALITION *for* RESOURCE RECOVERY

GLOBAL GREEN USA

New York City Waste Reduction Plan

State of the City Announcement to Double City's Waste Diversion by 2017

- Cost of waste export expected to increase by >40% in next 4 years.
- Supports PlaNYC greenhouses gas emissions reduction goals
- Supports solid waste diversion goals

COALITION *for* RESOURCE RECOVERY

GLOBAL GREEN USA

Key Points of New York City Waste Reduction Plan and Relevance of Post-Use Optimization of Packaging

- RFP for 450 tpd advanced conversion technology for MSW
- New MRF scheduled to be completed in first half of 2013 will accept all household rigids, contingent upon viable markets
- Initiatives to enhance composting infrastructure under development
- Full waste reduction plan will be announced in April 2012 with release of updated PlaNYC chapter

COALITION *for* RESOURCE RECOVERY

GLOBAL GREEN USA

Recyclable Wholesale Transfer Packaging

1.45 Million Tons of Waxed Boxes Discarded in the U.S. Each Year

COALITION *for* RESOURCE RECOVERY

GLOBAL GREEN USA

2010 - Largest U.S. Poultry Plant Converts to Recyclable Boxes

COALITION *for* RESOURCE RECOVERY

GLOBAL GREEN USA

The Produce Box Supply Chain

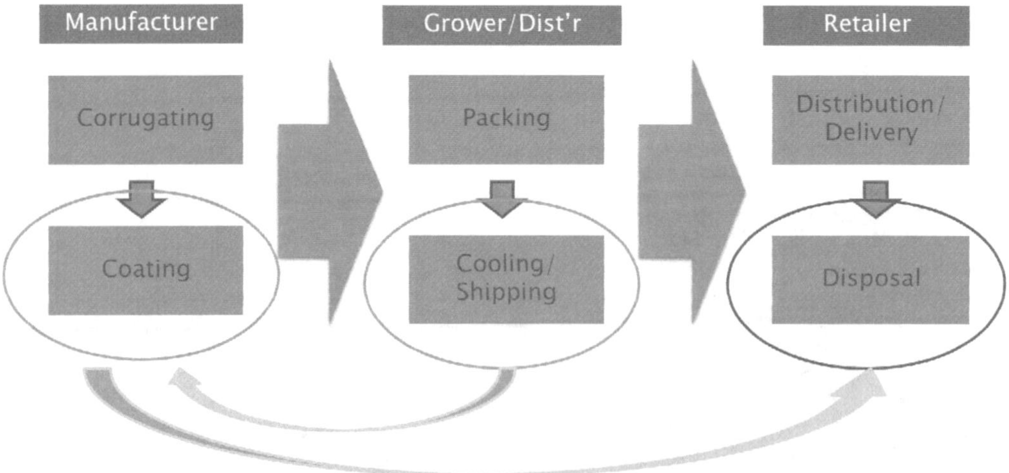

COALITION *for* RESOURCE RECOVERY

GLOBAL GREEN USA

Recyclable Packaging Demonstration & Press Event

April 19, 2012, Santa Cruz County, CA

- To **shine light on advances** in commercially available recyclable produce box products
- Common salad ingredients will be packed, cooled, and transported in recyclable wholesale packaging manufactured by multiple packaging companies from local farms to New Leaf Community Market's flagship store.
- Goal: Grocers requesting recyclable wholesale packaging as a replacement for wax coated boxes

GLOBAL GREEN USA COALITION *for* RESOURCE RECOVERY 4/20/2012

Projected Grocer Bottom Line Profit

Waxed Boxes per Day	25	50	100	250
Wax Boxes per Year	9,125	18,250	36,500	54,750
Cost to Landfill (tip fee)	$384	$ 768	$ 1,536	$ 2,304
OCC Lost Revenue	$913	$1,825	$ 3,650	$ 5,475
Potential bottom line profit	$1,296	$ 2,593	$ 5,186	$ 7,779
25 Grocery Stores	$32,412	$ 64,824	$ 129,648	$194,472
50 Grocery Stores	$64,824	$129,648	$ 259,296	$ 388,944
100 Grocery Stores	$129,648	$ 259,296	$ 518,592	$ 777,888
500 Grocery Stores	$648,240	$ 1,296,480	$2,592,960	$ 3,889,440
1,000 Grocery Stores	$1,296,480	$ 2,592,960	$5,185,920	$ 7,778,880

GLOBAL GREEN USA COALITION *for* RESOURCE RECOVERY

Global Green's Food Packaging Recycling Pilot Programs

GLOBAL GREEN USA COALITION *for* RESOURCE RECOVERY

Paper Foodservice Packaging Recycling Initiative Overarching Goal:

Recycle paper foodservice packaging into valuable, high-quality materials

Transforming Problem into Design Criteria

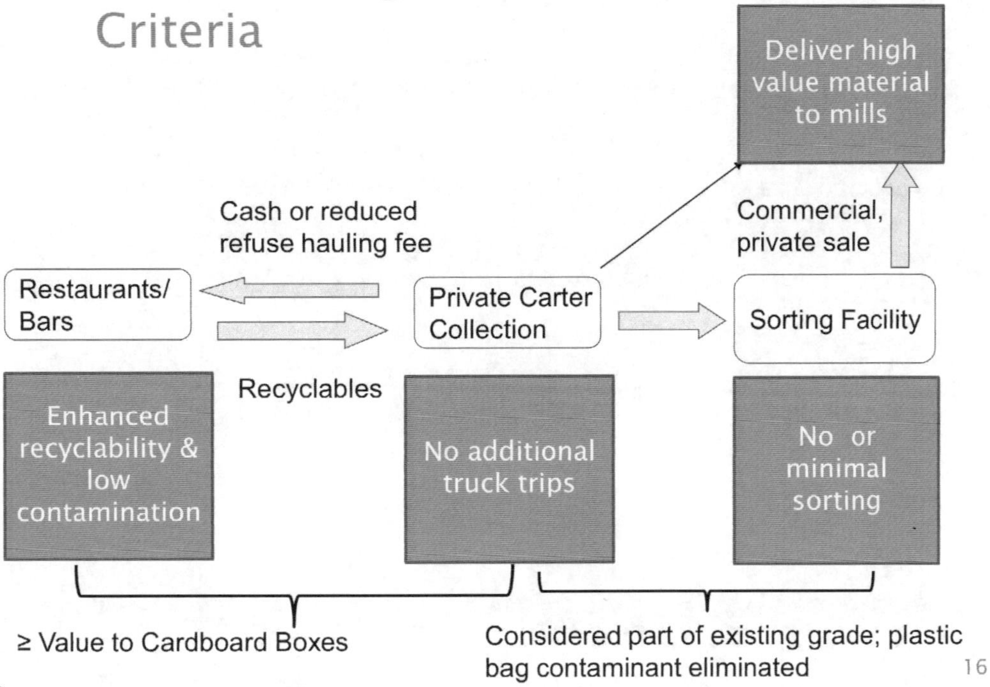

16

Material Pre-Screen

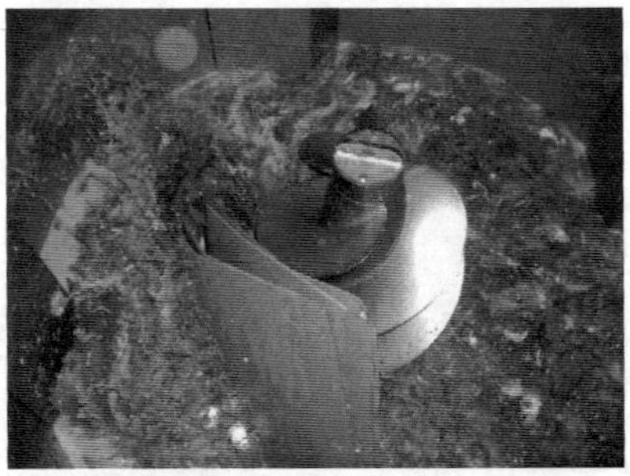

In-Store Design

Collect with OCC

Recycle

Summary of Key Findings

- ✓ Restaurants are interested in implementing recycling systems
- ✓ Customers will sort waste into specific streams, given sufficient cues
- ✓ The majority of paper foodservice packaging contains good fiber
- ✓ Making FDA-compliant, direct food contact paper packaging from post-consumer paper foodservice packaging Is possible
- ✓ Recyclable and compostable coatings are the game changer

COALITION *for* RESOURCE RECOVERY

GLOBAL GREEN USA

2012-2013

150 Store, 5 Brand Pilot in New York City

COALITION *for* RESOURCE RECOVERY

GLOBAL GREEN USA

Thank You

Annie White
Director, NY Office & Coalition
for Resource Recovery
Global Green USA
www.thecorr.org

COALITION *for* RESOURCE RECOVERY

GLOBAL GREEN USA

Closing the Loop: Removing Barriers to Material Recovery

Elizabeth Shoch
Sustainable Packaging Symposium
Houston, TX
April 5, 2012

SUSTAINABLE PACKAGING COALITION®
A PROJECT OF GREENBLUE

GreenBlue® is a nonprofit that equips business with the science and resources to make products more sustainable.

GREENBLUE®

The Sustainable Packaging Coalition is...

An industry working group dedicated to creating a robust environmental vision for packaging using a life cycle and supply chain approach...

consisting of over 200 member companies, educational institutions, and government agencies involved with the packaging supply chain.

SUSTAINABLE PACKAGING COALITION®
A PROJECT OF GREENBLUE®

Definition of Sustainable Packaging

Together, these criteria characterize our vision of sustainable packaging. No ranking is implied in the order of definition criteria.

- Is beneficial, safe & healthy for individuals and communities throughout its life cycle;
- Meets market criteria for both performance and cost;
- Is sourced, manufactured, transported, and recycled using renewable energy;
- Is effectively recovered and utilized in biological and/or industrial closed loop cycles;
- Is physically designed to optimize materials and energy;
- Optimizes the use of renewable or recycled source materials;
- Is manufactured using clean production technologies and best practices;
- Is made from materials healthy in all probable end of life scenarios.

closing the loop

- 3-year research grant from CalRecycle
- Systems approach (all packaging, all materials)
- Goal: Significantly increase the overall recovery of packaging by closely connecting packaging design with beneficial end-of-life options.
- Focus on best practices and infrastructure around the world: Collection, sorting, reprocessing equipment…

But what lies behind the infrastructure?

SUSTAINABLE PACKAGING COALITION®
A PROJECT OF GREENBLUE

Closing the Loop

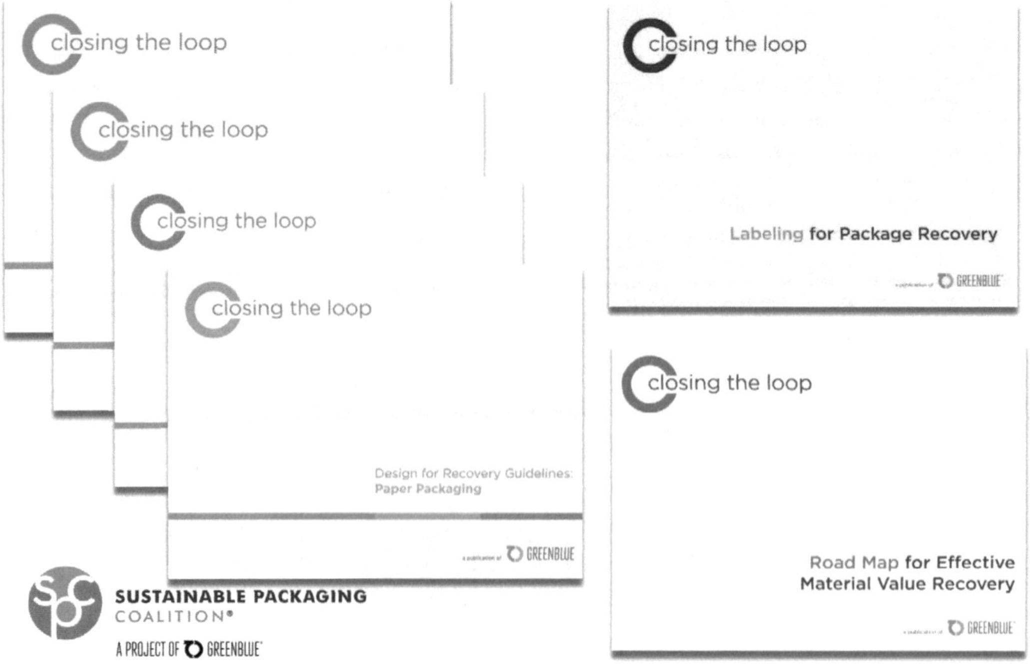

SUSTAINABLE PACKAGING COALITION®
A PROJECT OF GREENBLUE

Can We Take a Systems Approach?

- All packaging materials
- All package formats
- All locations
- Swiss model: each material association is responsible for collection and recycling of its material

SUSTAINABLE PACKAGING COALITION®
A PROJECT OF GREENBLUE

Systems Approach: Recycling in Rural Areas

- 21% of population in US (2000 US Census)
- Lower population density, lower volume of material
- Geography of people, MRFs, reprocessors
  - Efficiency of routes
- Financial limitations
- Availability of bins or drop-off sites

SUSTAINABLE PACKAGING COALITION®
A PROJECT OF GREENBLUE

What is the Goal?

- Diversion from landfill?
- Recycling?
- Development of local secondary markets?
- Commitments to use recycled content in packaging?
- Two or more of the above?

SUSTAINABLE PACKAGING COALITION®
A PROJECT OF GREENBLUE

Policies at Cross-Purposes

- Ambitious recycling goals but low landfill tipping fees
  - UK: annually increasing landfill tax
- State or local laws discourage siting of MRFs, composting facilities, WTE plants
- Escheats of funds that go to unrelated budget line items

SUSTAINABLE PACKAGING COALITION®
A PROJECT OF GREENBLUE

The Big Question: Funding

- US states and municipalities are hard hit by economic downturn
- All material recovery methods cost money
- Role of local & state government, industry, other?

Headline: "Trash fees will be lower"

SUSTAINABLE PACKAGING COALITION®
A PROJECT OF GREENBLUE®

Infrastructure: Curbside Bins

- Four or five bins for best quality (and price, and demand…)
  - Glass, paper, other recycling, organics, residual
  - At a minimum, separate glass
- Drop-offs
- What about on-the-go bins?

SUSTAINABLE PACKAGING COALITION®
A PROJECT OF GREENBLUE®

R6R 042931
R6B0707 924693

TRAVEL

Flaps open when pushed
Litter
Recycling
Recycling
Litter

Which Materials are Collected? How?

- Standardization of materials collected across jurisdictions
  - Efficiencies in collection, sorting, marketing of scrap
  - Easy for public to understand
- Not everything can be recycled economically
  - Decide what makes sense to recycle, and focus on collecting those packages
  - BUT - find a beneficial end-of-life option for all packaging (WTE, anaerobic digestion and composting)

SUSTAINABLE PACKAGING COALITION®
A PROJECT OF GREENBLUE®

What's Next? New Technologies

- State of the art MRF sorting technology
  - Optical sorting
  - Remove films
- Market development: where will these collected and sorted materials go?

SPC COALITION®
A PROJECT OF GREENBLUE®

Waste to Energy

- WTE not built for energy, but for waste disposal.
- What about recycling vs incineration?
  - EU countries have two Packaging Directive targets: recycling and recovery

SUSTAINABLE PACKAGING COALITION®
A PROJECT OF GREENBLUE®

MVA Pfaffenau – Vienna, Austria

- Waste disposal: 250,000 tonnes/year
- Electricity: 70 GWh/year
- Heat: 400 GWh/year district heat
- Cost of plant: €200 million (built 2008)
- Estimated revenue from heat & electricity: €8 million/year
- Financing 4.5% over 20 years = €15 million/year
- "…since you also have costs for plant operation and maintenance, you can see it is not possible to finance a typical waste incineration plant with high environmental standards only by energy production."

SUSTAINABLE PACKAGING COALITION®
A PROJECT OF GREENBLUE®

Public education

- On-going need, even in systems with high recycling rates
- Consider additional languages, cultures
- Areas with tourists or seasonal residents
- Create programs in schools

SUSTAINABLE PACKAGING COALITION®
A PROJECT OF GREENBLUE

Label – The Consumer's Role

- In 2012, SPC members introduced a new recycling label into the marketplace
- Use is voluntary!
- Labels all package components, all materials, nationwide
- Recyclability determination supported by access-to-recycling data

SUSTAINABLE PACKAGING COALITION®
A PROJECT OF GREENBLUE

To remove barriers to recycling in the US…

1. Take a systems approach: consider all materials, all packaging formats, and all end-of-life options.
2. Start by determining the system's goal: recovery of materials, quality of materials, environmental benefit, diversion from landfill, targets, development of local secondary markets, etc.
3. Ensure underlying policies are coordinated and not working against each other, creating conflicting incentives.
4. A sustainable financing model is needed to alleviate the financial burden on local governments.
5. For best material quality and value, a four-bin system (or more) is preferred.

SUSTAINABLE PACKAGING COALITION®
A PROJECT OF GREENBLUE®

To remove barriers to recycling in the US…

6. Standardize the materials collected and collection method across a state or country so collection and sorting infrastructure can also be standardized. Collect and sort only the materials that make sense economically
7. Invest in new sorting technology (including optical sorting technology) at MRFs and in market development for recycled materials.
8. Consider anaerobic digestion and industrial composting for food waste and compostable packaging and re-evaluate waste-to-energy technologies for materials that are not economically recyclable and for which landfill is currently the only option.

SUSTAINABLE PACKAGING COALITION®
A PROJECT OF GREENBLUE®

To remove barriers to recycling in the US…

9. Collect state-wide data on access to recycling as well as rate of recycling for all materials and packaging formats to ensure targets are reached and enforceable.
10. Consumer education is an on-going requirement for success. A good recovery label can help.

SUSTAINABLE PACKAGING COALITION®
A PROJECT OF GREENBLUE®

SUSTAINABLE PACKAGING COALITION®
A PROJECT OF GREENBLUE®

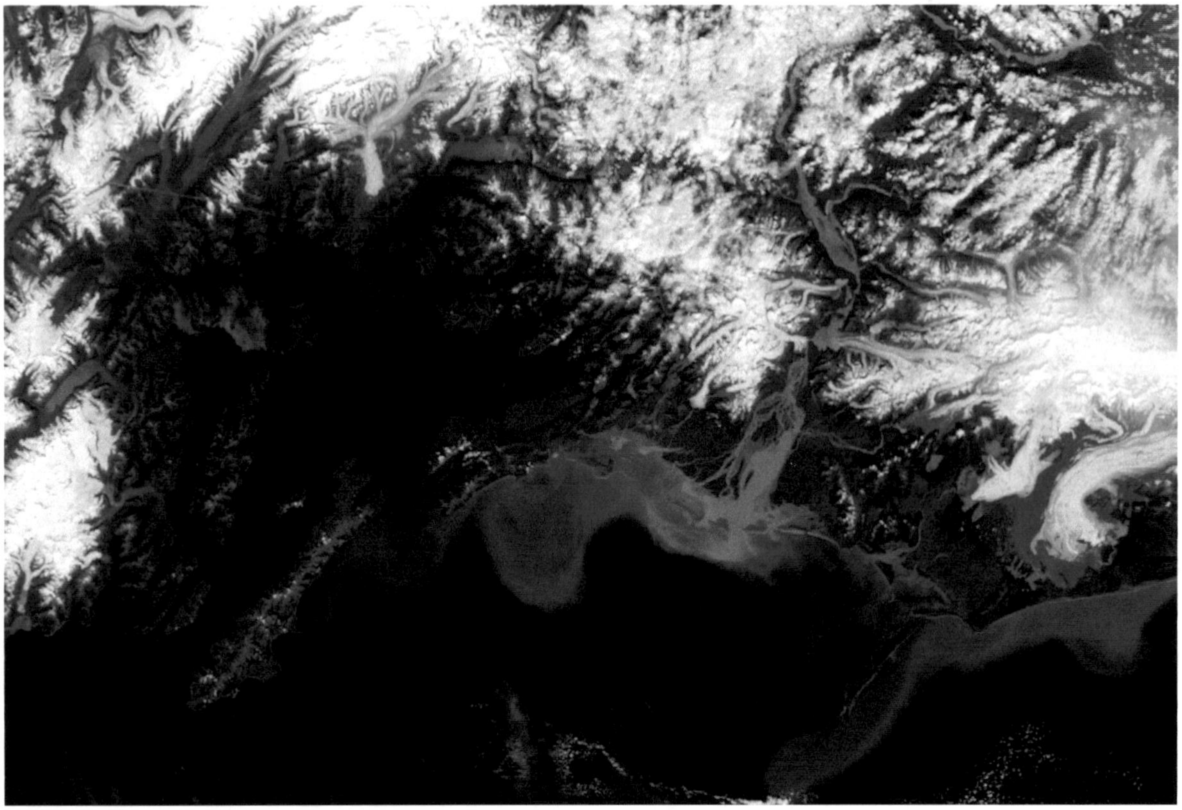

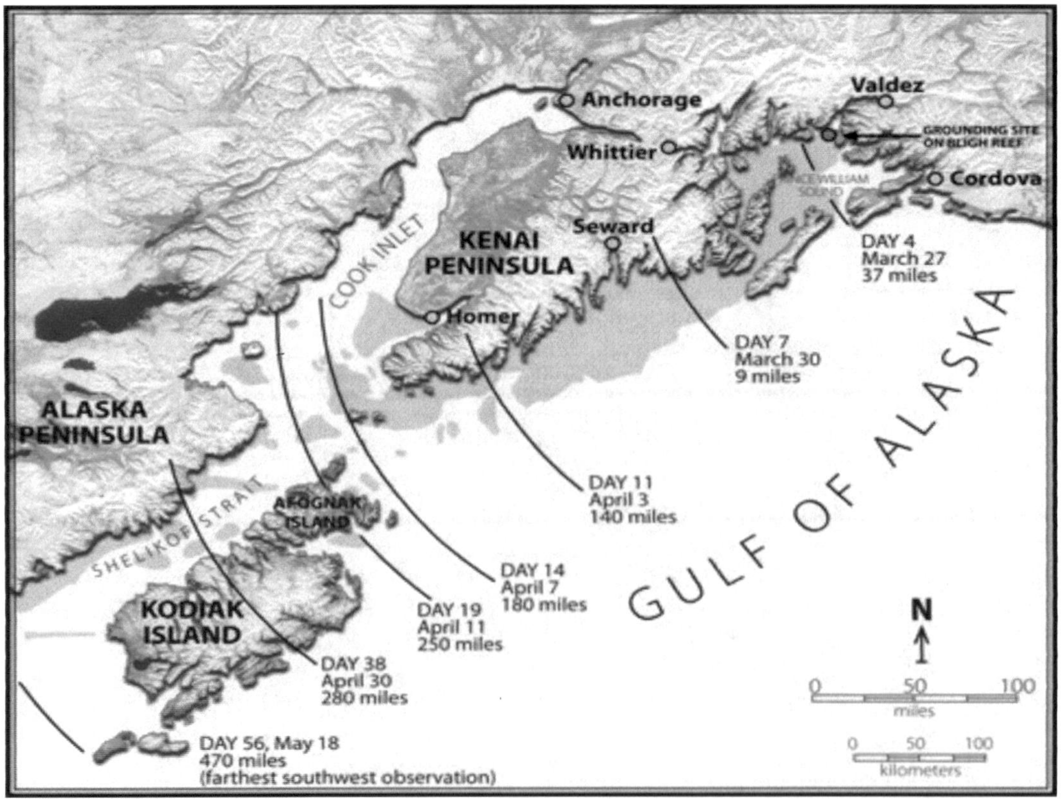
Anchorage
Valdez
Whittier
GROUNDING SITE ON BLIGH REEF
Cordova
Seward
KENAI PENINSULA
COOK INLET
DAY 4
March 27
37 miles
Homer
DAY 7
March 30
9 miles
ALASKA PENINSULA
GULF OF ALASKA
DAY 11
April 3
140 miles
AFOGNAK ISLAND
SHELIKOF STRAIT
DAY 14
April 7
180 miles
KODIAK ISLAND
DAY 19
April 11
250 miles
N
DAY 38
April 30
280 miles
0 50 100
miles
DAY 56, May 18
470 miles
(farthest southwest observation)
0 50 100
kilometers

Crossing the
Copper River Delta

Bering Lake, West Side

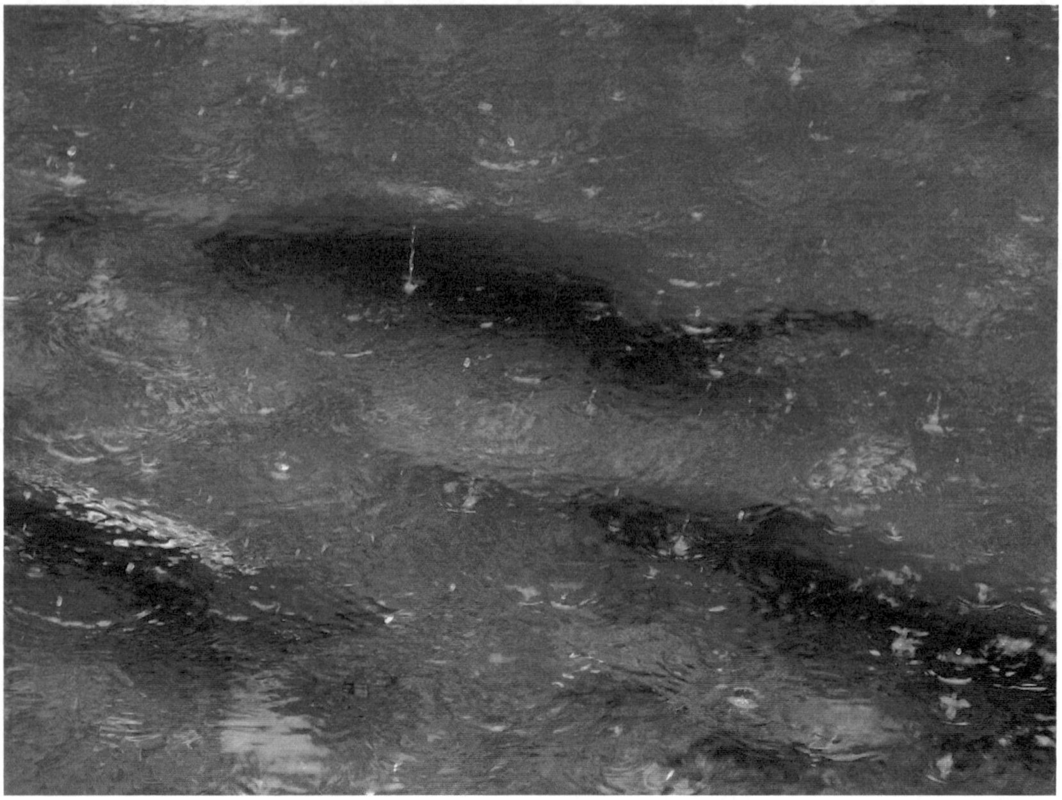

IF YOU MUST SMOKE

SEAFOOD CHAMPION
DUNE LANKARD

travenous
ssure Bleed
Control

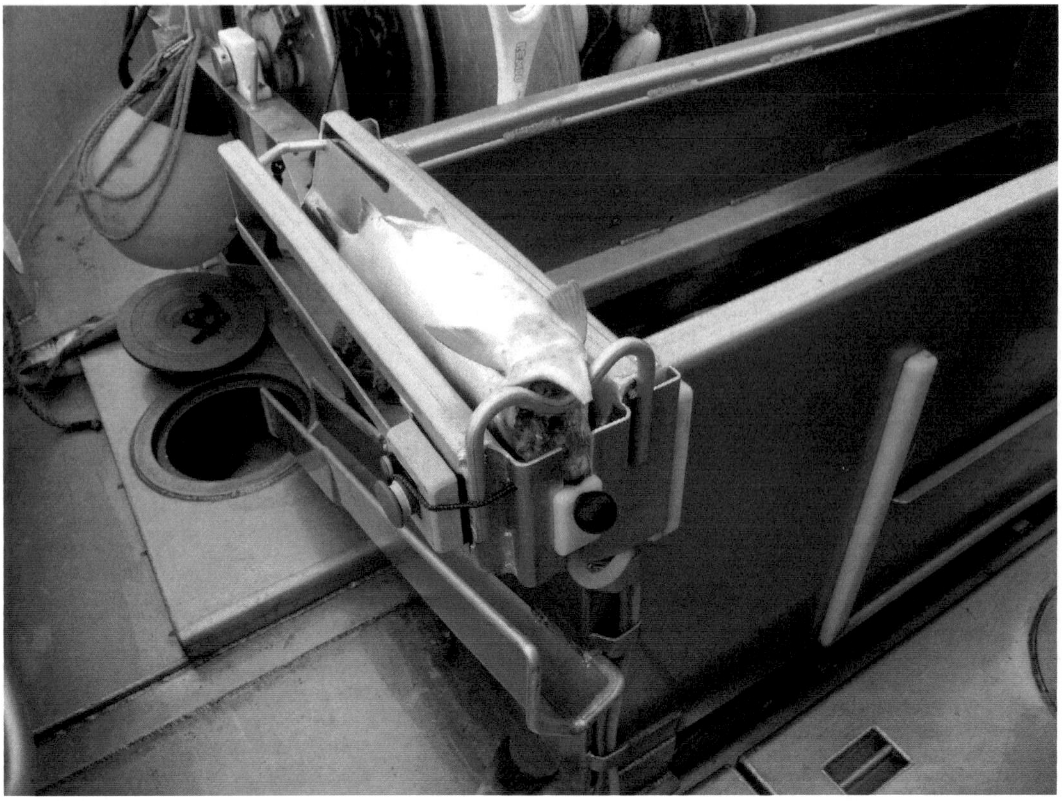

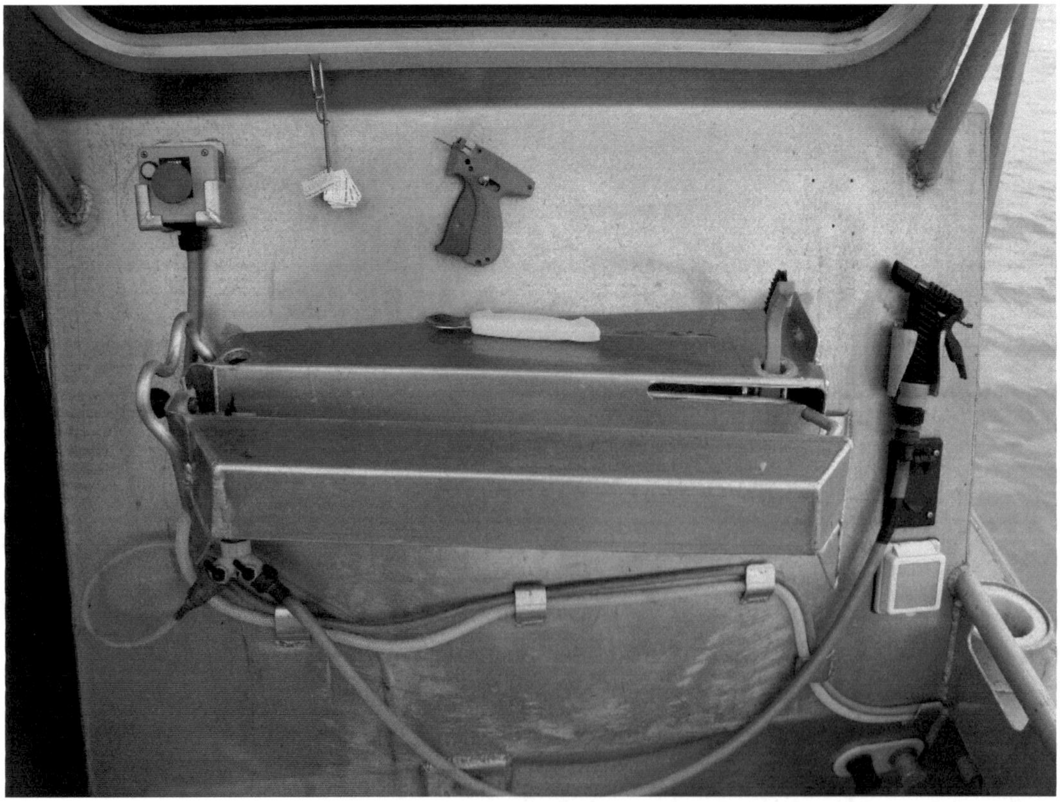

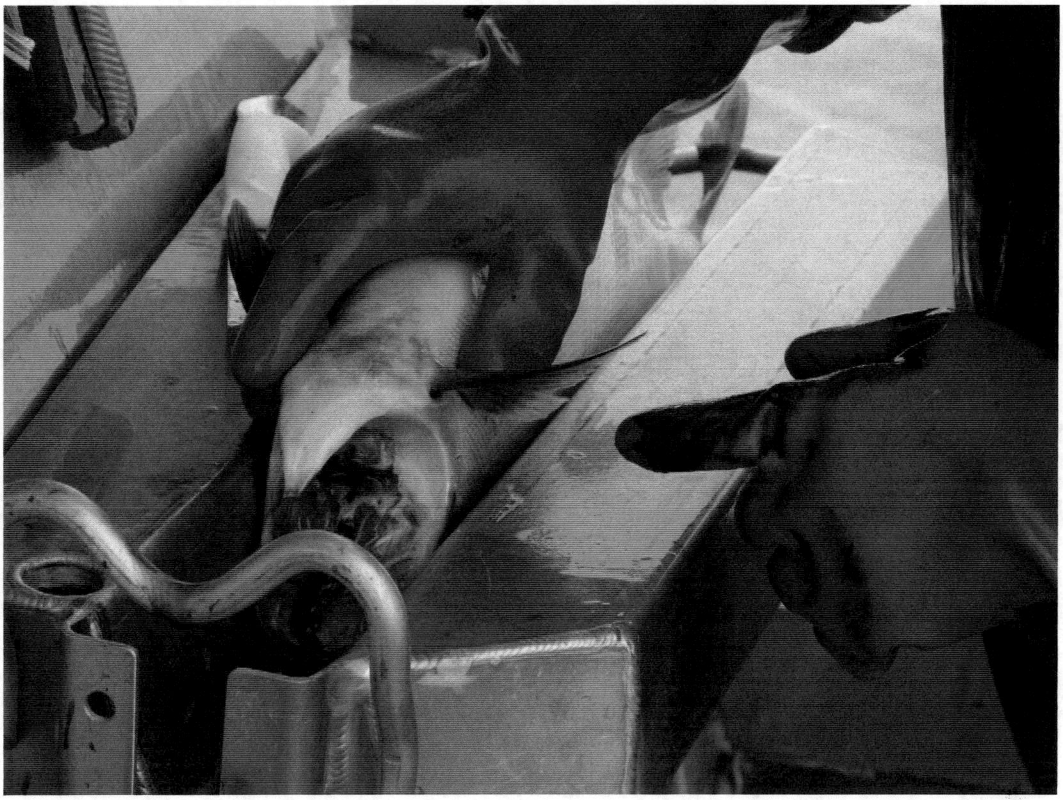

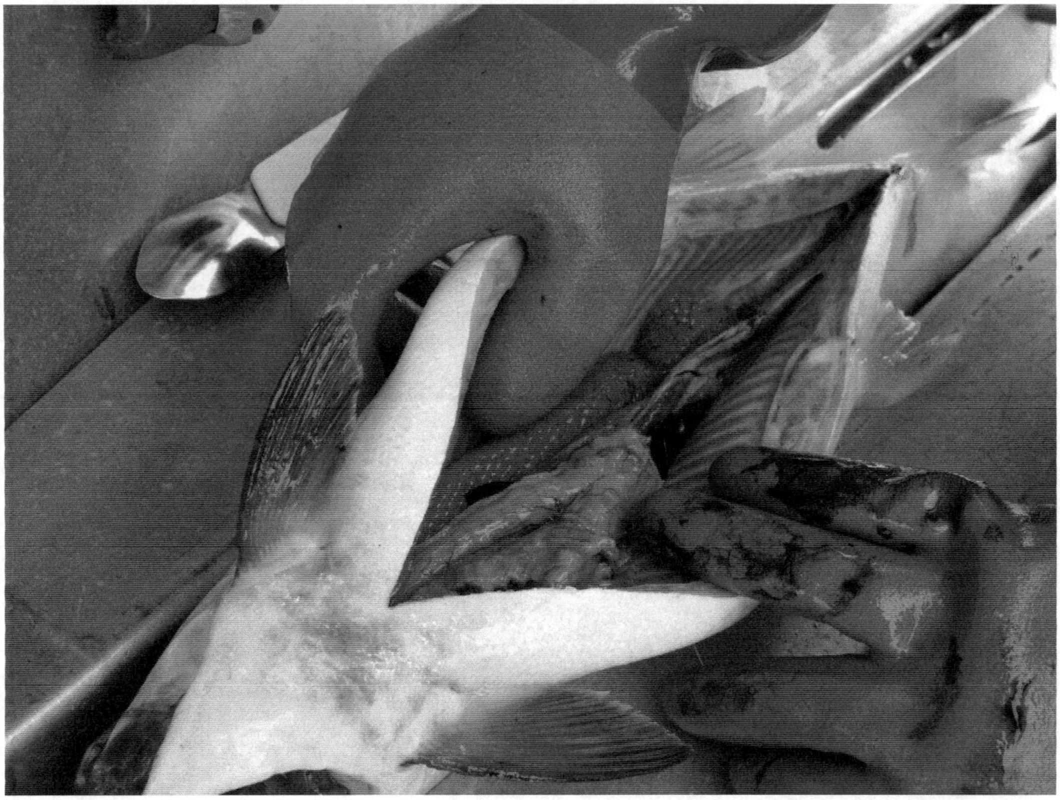

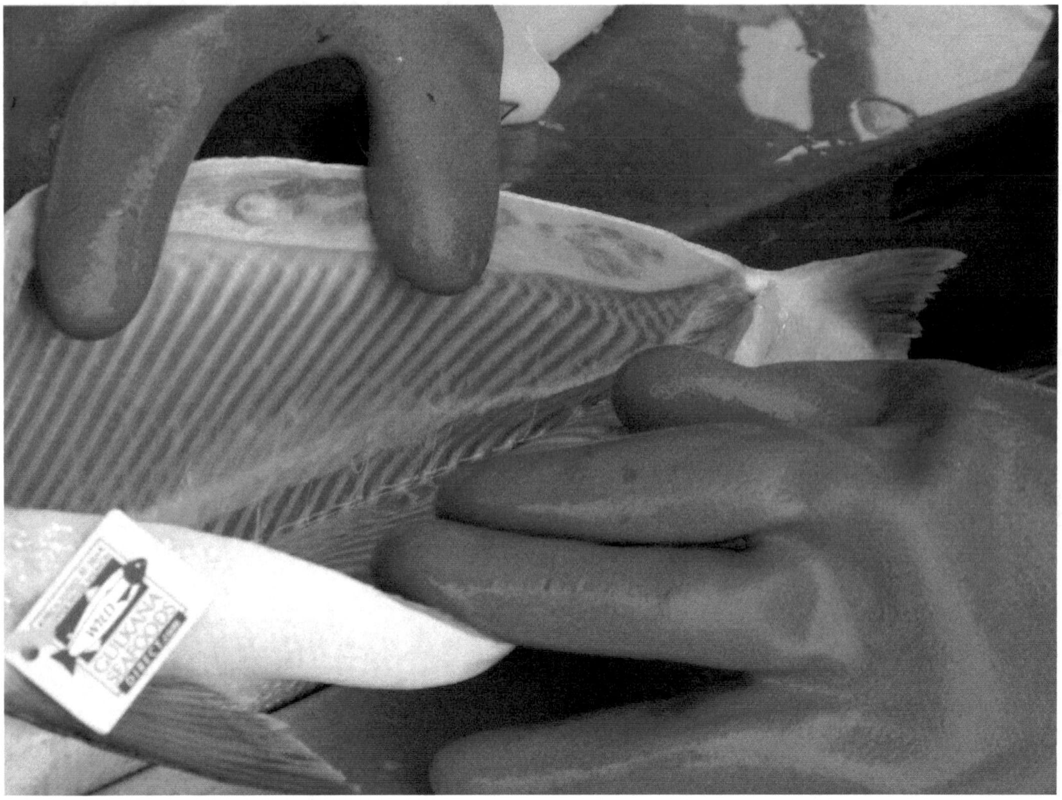

CAROL HOOVER
SFO
PRI FRO
Alaska Airlines / Horizon Air
FROZEN
THIS SIDE UP
Wild Fish PRODUCT OF U.S.A.

GEL - ICE PAK
NON TOXIC - RINSE ANY LEAKAGE WITH WATER
GEL - ICE PAK
NON TOXIC - RINSE ANY LEAKAGE WITH WATER
FROZEN
THIS SIDE UP
Wild Fish PRODUCT OF U.S.A.

FROZEN
THIS SIDE UP
Wild Fish PRODUCT OF U.S.A.

dune@redzone.org
or
eyak@redzone.org
Thank you

AUTHOR INDEX